WILD HONEY BEES

An Intimate Portrait

With photography by INGO ARNDT
and text by JÜRGEN TAUTZ

PRINCETON UNIVERSITY PRESS
Princeton and Oxford

ABOUT THE AUTHORS

Ingo Arndt is one of the world's top nature photographers. His work appears in international magazines and newspapers such as *GEO*, *Stern*, *National Geographic* and *BBC Wildlife*. His photographs have won numerous prizes, including a World Press Photo Award. He has won the Wildlife Photographer of the Year Award several times.

Prof. Dr Jürgen Tautz is a behavioural scientist, sociobiologist and bee expert. He is a Professor at the Biocenter of the Biozentrum der Julius-Maximilians-Universität in Würzburg and chairman of its bee research group. Tautz is the author of several popular science books on bees and holder of the Communicator Award of the DFG (Deutsche Forschungsgemeinschaft – German Research Foundation).

PAGES 8/9
Honey bees can obtain important information about their surroundings using their compound eyes, feelers and countless sensory hairs in order to carry out their day-to-day tasks.

—

PAGES 10/11
Worker bees loading up with water needed to cool their colony's nest on a hot day.

—

PAGES 12/13
Inspecting an empty cell clarifies what needs to be done.

—

PAGES 14/15
View of freshly made combs in a honey bee nest inside a tree cavity.

—

PAGES 16/17
Bee nest in an abandoned black woodpecker nest, high above the forest floor.

I THINK EVERYONE NEEDS HONEY BEES, at least a little. Watching them bobbing on flowers as they collect their nectar and pollen, or gathering water at the edge of a small stream, or zipping in and out of their fortress-like homes, can give us a daily measure of wonder and a glimpse of how life goes on beyond our human affairs. This is especially true for the colonies of honey bees that live wild and free, nesting in hollow trees hidden deep in forests. The behaviour, social life and ecology of these free-living honey bees is the subject of the marvellous book that you now hold in your hands. Behavioural biologist Jürgen Tautz and wildlife photographer Ingo Arndt have teamed up to take readers on a stunning visual journey that reveals with unsurpassed beauty the natural history of *Apis mellifera*, our greatest friend among the insects.

Thomas D. Seeley
Professor in Biology, Department of Neurobiology and Behavior, Cornell University

180
EPILOGUE

184
BEE PHOTOGRAPHY

187
BIBLIOGRAPHY

188
THANK YOU

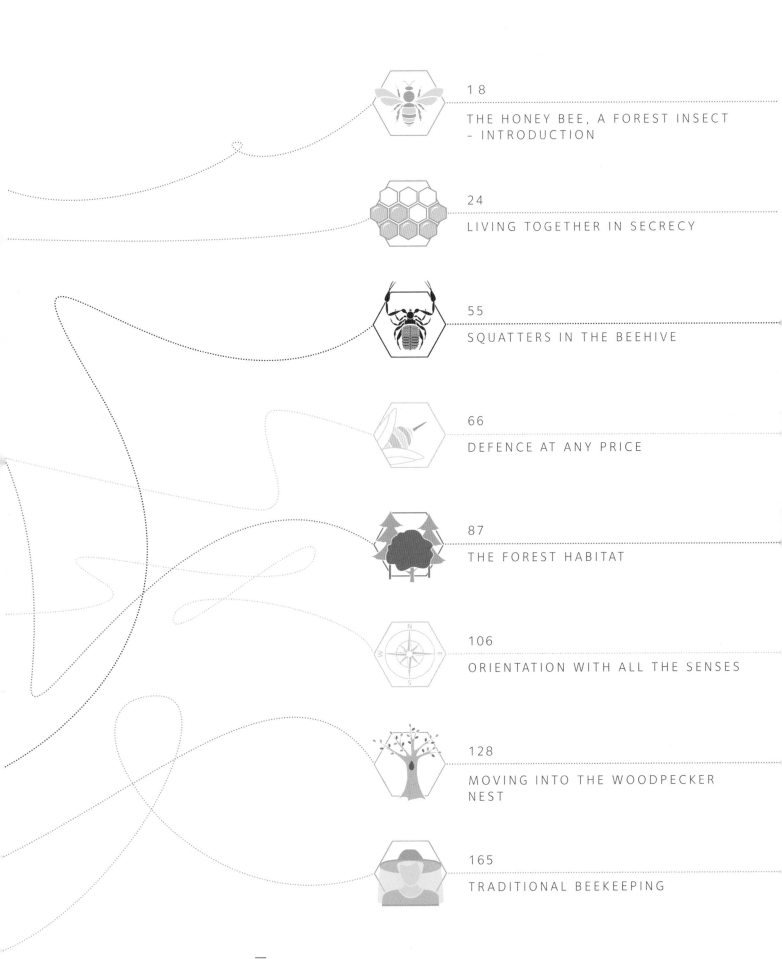

18

THE HONEY BEE, A FOREST INSECT
– INTRODUCTION

24

LIVING TOGETHER IN SECRECY

55

SQUATTERS IN THE BEEHIVE

66

DEFENCE AT ANY PRICE

87

THE FOREST HABITAT

106

ORIENTATION WITH ALL THE SENSES

128

MOVING INTO THE WOODPECKER
NEST

165

TRADITIONAL BEEKEEPING

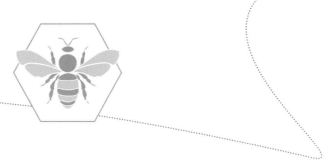

THE HONEY BEE, A FOREST INSECT

INTRODUCTION

The cold season stops most insects in their tracks. They survive chilled and in a state of suspended animation under loose tree bark or buried in the ground. They survive as eggs, larvae, in the pupal stage, or even as fully developed insects, like the future queens of wasps, hornets or bumble bees, the only members of their species not to die in the autumn.

In contrast, colonies of honey bees spend the winter well protected and, by generating their own heat, snugly tucked into hollow trees in the forest. At least that was their natural habitat until humans began transforming the world at large and that of the bees with it. Ever since beekeepers began housing honey bees in hives, these insects have been known mainly as ones that live in hives.

But wild, forest-dwelling honey bees do still exist in hollow trees – even in temperate latitudes. And they are more important and valuable than ever before, especially now when the significance of preserving the biodiversity in our forests is more widely understood, and knowledge of the biology of wild bees could help to reshape the practice of beekeeping.

We set out on the trail of wild honey bees, observing and photographing their behaviour, and in doing so obtaining insights into hitherto unknown details of their way of life. During our research we were out and about in forests, where we encountered *Apis mellifera*, the western honey bee. In addition to this species, there are many other species of bees around the world, of which the eastern or Asiatic honey bee is the most important, occurring primarily in Asia and being responsible for transmitting a dangerous parasite to our bee colonies – the *Varroa* mite.

A remarkable number of wild honey bee colonies still exist, living in hollow trees inside the forest, largely unnoticed by humans and in greatly underestimated numbers. They live in conditions that have shaped their characteristics and abilities across countless generations. Therefore, given everything we know today, it should come as no surprise that these honey bees, original dwellers of the forest, are better able to cope with diseases and parasites than our 'pet' bees. They have no choice in the matter as there is no beekeeper there to help them. Nature provides them with what they need to survive and multiply.

If one takes a closer look at honey bees in the forest, in their ancestral habitat, it quickly becomes clear that they are deeply integrated in highly complex relationships and play a paramount role in the preservation and structure of their environment.

Habitats are especially stable when they are based on a variety of species of flora, fauna and microorganisms, which are all interlinked and interdependent. The forest is no exception.

A bee collecting honeydew on a fir tree, which will be converted into forest honey back in the bee's nest.

But what is a forest? What distinguishes it? It cannot exist without trees, but a bunch of trees does not constitute a forest. It is essential that the trees stand close together, in contrast to an open, wooded park landscape. The tightly packed trees create a forest climate, and thus a characteristic that is considered entirely unique. A forest climate is characterized by steady temperatures, moderate air movements, dampened light and high humidity. These factors are not only climatological parameters but also living conditions, for the honey bees as well. Many of the attributes and abilities inherent in honey bees can be explained by their life as forest dwellers. Indeed, many of the problems affecting honey bees housed in hives today can be understood by regarding the forest as their natural habitat and the one to which they are adapted.

»MANY OF THE ATTRIBUTES AND ABILITIES INHERENT IN HONEY BEES CAN BE EXPLAINED BY THE LIVING CONDITIONS IN THE FOREST.«

—

OPPOSITE
Forager bee preparing to land on a goldenrod.

—

OVERLEAF
Wild garlic, a flowering paradise for forest-dwelling bees.

We humans have drastically interfered with the natural life of honey bees. We keep them in geometrically exact, standardized boxes, made of wood at best. We have radically changed our environment and that of the honey bees. Far-reaching, contiguous woodlands have disappeared and, where they still exist, they are rarely used as homes for bee colonies. Commercially optimized forestry has led to forests being clearly laid out as delineated monocultures in order to simplify processing, and to them no longer bearing any resemblance to the healthy, well-balanced forests that once existed.

Such habitats are especially vulnerable to parasites as well as to the rapid warming of our atmosphere, which we refer to as climate change. This should give us pause to rethink what type of forest we actually want. It is fascinating to see that the honey bee is a central subject in this context as well. However, it becomes clear straight away how little we know about the life of honey bees in their natural forest environment.

The forest has ceased to be the natural habitat for the vast majority of bee colonies that exist today. To a great extent it has been replaced by cultivated land, thus causing changes in the home environment to impact 'pet honey bees' even more severely. Tree cavities host a wide array of other living creatures, representing a 'mini biotope', which is not without impact on the bee colonies, and vice versa.

Even the physical conditions offered to the bees inside the tree are not necessarily found in the structures used by beekeepers to house their bee colonies. This does not necessarily pose a problem in itself – after all, honey bees are extremely resilient creatures – but we do not yet have sufficient knowledge about the complex interrelationships to draw conclusions about whether or not the unintended departure of bees from the forest creates problems for the animals.

What we can say for certain is this – wild honey bees are a natural element of a healthy forest. Both realities should coexist – bee colonies cared for by beekeepers for our mutual benefit, and wild bee colonies which constitute an important component of the forest ecosystem. Beekeepers can offer their honey bees the best possible living conditions and can at the same time profit from the natural selection that takes effect among wild bee colonies.

This book is intended to provide a closer look at the life of honey bees inside the forests of Central Europe. With glimpses into their lives that have never been seen before, we hope to contribute to honey bees being seen for what they truly are – creatures of the forest.

Ingo Arndt & Jürgen Tautz

LIVING TOGETHER IN SECRECY

Cavities provide shelter from the hardships of the world outside. Countless animals use them, taking advantage of ones that already exist or creating their own. Our ancestors rode out harsh winters inside caves, surviving attacks from predators or hostile groups of people. Being cavity dwellers provides honey bees with significant protection against the weather and their enemies. Honeycombs built out in the open without protection would result in the bee colony freezing to death in the winter, while birds and other animals could easily steal their honey and make short work of the bee brood in the summer.

Cavities are an indispensable component of the bee lifestyle, and if, as is the case with the honey bee, bees cannot create them for themselves, they are forced to rely on a dependable 'housing supply' from elsewhere. It is impossible for the bee population to grow without a sufficient supply of cavities offering bee-friendly features. At this point, it becomes clear how closely interconnected all life is and this holds true for Central European forests. In the case of bee dwellings, woodpeckers drill holes into trees for their own nests. Once the woodpecker holes are no longer being used by their builders, they are available to a wide range of other interested parties, including honey bees.

The chapter 'Moving into the woodpecker nest' is devoted to the process of honey bees moving into a hollow trunk, setting up the honeycombs, and the steps they take to adapt the cavity to suit their needs. The set-up phase, i.e. the time from which the hollow is colonized until the completion of the honeycombs in the first year, is a highly critical stage that determines whether or not the colony can gain a substantial foothold. Most bee swarms do not make it. We don't know for sure about the exact reasons for swarm losses during the foundation phase. Were no suitable cavities available? Did the swarm not have enough manpower to procure the nectar and pollen required to lay a brood? Were there even enough sources of food available at the right time and within close enough vicinity to the new nest? However, once the new construction has been completed successfully, the daily routine of the bee colony sets in, provided no misfortune befalls the bees.

The honeycombs form the centre of life inside the honey bee hollow. They serve as production facilities and storage space for honey, a repository for pollen, a nursery for larvae

—

A forager bee fluttering into the newly colonized tree cavity early in the morning. It is still too cold outside for her hive-mates. It is not the light at the break of dawn but the air temperature at the entrance of the nest that serves as a start signal for the foraging flights. Fanning bees around the entrance see to the air circulation.

and pupae, and a 'phone network' for communication. The combs feature chemical markers, which act as signposts for the bees in the dark nest and assist them in extracting optimal use of the wax structures. The way in which they use the individual comb cells is impressively well-ordered. There is no confusion, but rather a hive, divided in an orderly manner into sections for producing and storing honey, storing pollen, and rearing larvae and pupae.

Taking a look at this order inevitably begs the question: who directs the activities of all the bees with such impressively organized results? The answer – nobody. All emergent results, arising from the activities of thousands of bees, are the consequence of the individual decisions made by the worker bees. Each bee decides for itself what to do. The grounds for its decisions are manifold and can be whatever influences it at that particular moment in its immediate vicinity. It could be stimuli such as temperature, or scents and communication signals from other hive-mates, whatever it has retained in its memory based on its own experience, and its motivation to perform or not perform a certain action. The genetic material bees inherit has a strong influence on how sensitive they are, and thus how quickly they react. The multiple mating flights of the virgin queen during her nuptial flight result in numerous half-sister lineages within the bee colony and thus very different 'characters' among the worker bees, although they might all appear the same on the outside. This makes

it possible, for the bee colony as a whole, to always have the right response to external disturbances.

One variable in the nest that has to be regulated is the temperature. A temperature of around 35 degrees Centigrade (95 degrees Fahrenheit) is established by the heater bees in the brood nest in order to provide the pupae with the optimal level of warmth. Astonishingly, the developmental stage before the emergence of an adult bee, during which the brain is fully developed, requires a temperature very close to that of the human body. For us as well, our brain is the organ requiring this high temperature in order to operate properly. Heater bees, tucked away in empty cells inside the brood nest, deliberately direct their warmth towards neighbouring pupae. While doing so, a single heater bee can supply heat to up to 70 pupae on both sides of the comb. Heater bees sitting on a honeycomb and pressing their heated thorax against a cell cap in the brood nest (acting as a nursery for a dormant pupa) emit heat into the ambient air in particular. However, that heat is not lost, as it serves to regulate the state of equilibrium, the so-called homeostasis, inside the bee nest. This heat is also essential for managing the air humidity in the cavity.

In a bee nest inside a hollow tree, the air does not form uniform bodies, or air masses, but rather is divided into streams of air by the combs. Each stream of air contains its own microclimate, which is controlled by the bees. They

OVERLEAF
Worker bees interlock their legs to form a net around the combs. The bees fashion these three-dimensional structures at night in particular. The 'bag' can feature a wide mesh or it can be pulled tightly together. In beekeeping hives and observation hives, this netting is severely limited, or not possible at all. Nevertheless, honey bees housed in hives form what are known as construction chains while building the combs, when there is more space in the hive, with multiple insects joining together and hanging under the combs like a chain. It was assumed that the bees did this in order to construct the combs vertically. However, it is possible that this behaviour is completely unrelated to the construction and instead can be ascribed to the original net building activities of wild bees.

generate warmth, especially for the capped pupae in the brood nest, and they can produce evaporative cooling by bringing in water and then creating air currents by beating their wings whilst remaining stationary. The bees' activity has a strong impact on the air temperature and thus on the air humidity in the alleys of the comb. Maintaining a low humidity in the bee nest has many advantages. The most obvious benefit is that it reduces the workload when thickening the nectar to turn it into honey, which becomes easier as the ambient air becomes drier. Furthermore, low air humidity prevents mould from forming.

How does a bee dehumidifier work? Due to the small opening that most tree hollows housing bees have, it is not possible to exchange air on a large scale no matter how hard the tiny, living ventilators might try. But just a few worker bees, positioned at appropriate locations atop the combs and generating an air current with their wings, can do this very effectively. They can direct cooler air from farther below out of the hollow and up between the combs. There, the air is heated, thus reducing the relative humidity. However, this does not diminish the mass of water vapour as the water does not simply disappear. What may help reduce the water vapour is that, according to recent studies, the coating honey bees put on the walls of their nest cavities has micropores in it that allow water vapour to escape.

»IF ONE OBSERVES THE HONEY BEE, ONE QUESTION ARISES IMMEDIATELY – WHO DIRECTS ALL OF THE ACTIVITIES THAT CONTRIBUTE TO THIS APPARENT SYSTEM OF ORDER?«

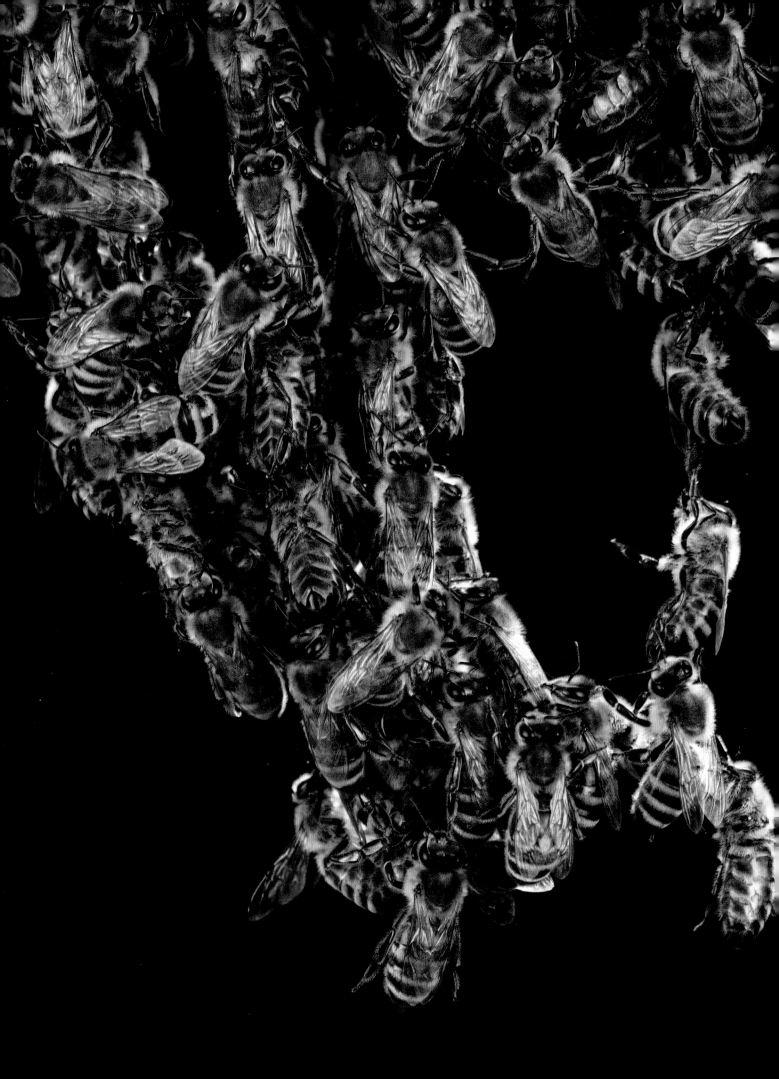

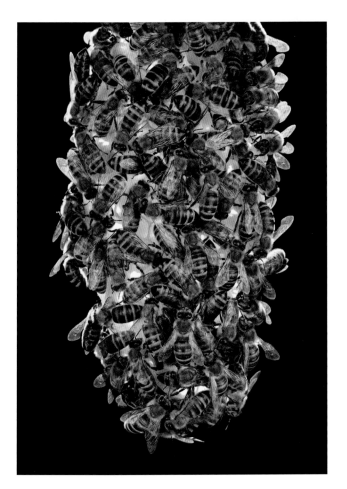

—

OPPOSITE

The original drop-shaped comb is expanded substantially within just a few days. To do this, the bees repeatedly create new saucers in tightly packed spheres, then stretch them out into tubes and warm up the wax. The internal stresses in the wax cause the walls separating one cell from its six neighbouring cells to organize themselves into homogeneous sections on their own, forming the perfect, hexagonal cell structure.

—

ABOVE

The extensive operations on the wax construction site are hardly perceptible with the wax-sweating and comb-building bees sitting so closely together (left). The growing combs only gradually come into view (right).

»THEIR CONSTRUCTIONS ARE ENDURING, CAN BE USED ACROSS MANY GENERATIONS, AND ARE EVEN DECOMPOSABLE.«

—

ABOVE

While visiting flowers, pollen gets caught on the fur' of the foragers and is brushed out with their legs when flying back home, clinging to the back legs in a small package. The colour of the pollen indicates which flowers were visited (left).

—

OPPOSITE AND OVERLEAF

The pollen is pressed tightly into empty cells and stored. Mixed with a bit of honey to create what is known as bee bread, this serves as a valuable nutritional source for the colony.

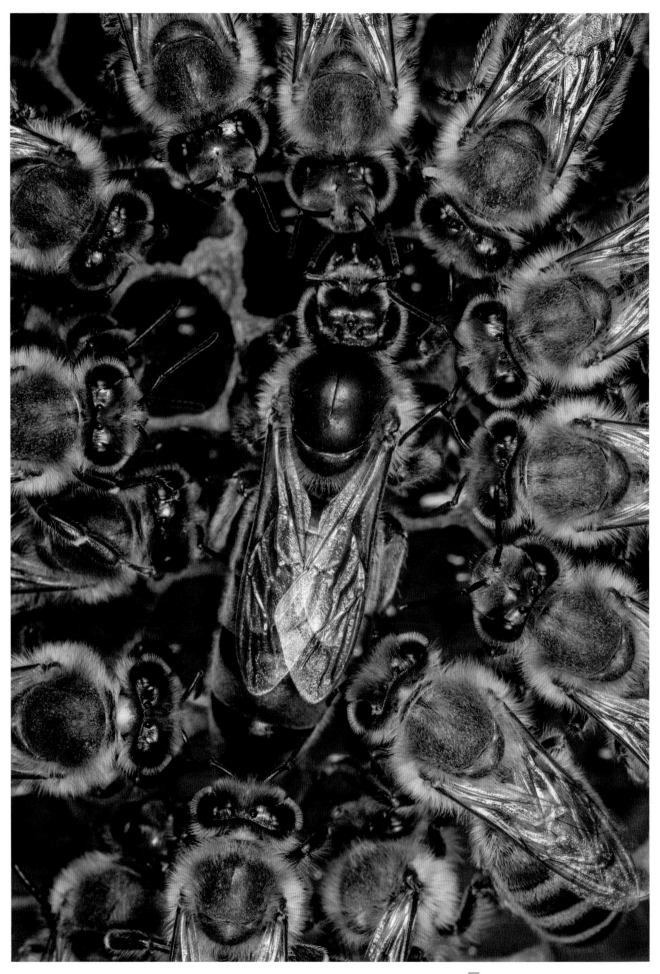

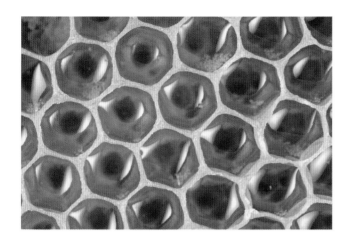

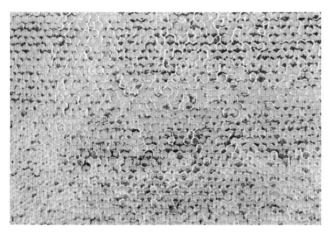

—

OPPOSITE

Queen with her court: worker bees which constantly feed and clean her, and even remove her faeces.

—

ABOVE AND RIGHT

Fresh nectar glistening in the cells (top left). The bees remove water from it, thus turning it into honey, and then cap the cells (bottom left). Bees inside the hive accept the honey from worker bees outside and store it in the combs (top right). A bee, possibly an exhausted heater bee, begging for honey from a dispenser bee (bottom right).

—

OVERLEAF

Queen laying eggs. Court bees escort the queen as she chooses an empty cell – 'parking in reverse'.

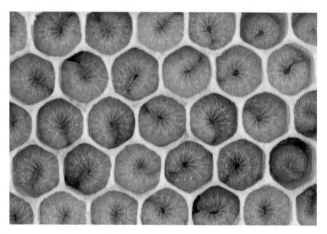

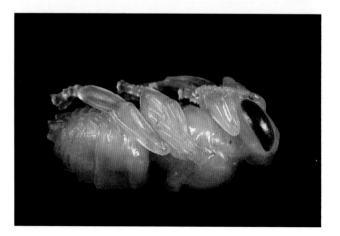

LEFT, FROM TOP TO BOTTOM
The eggs are secured individually to the bottom of each cell.

Before the final pupation, the larvae barely fit in their nursery anymore.

The metamorphosis into a fully developed bee occurs in the pupal stage. The pupae rest on their backs in their sealed cells. The cell cover was sliced off for this shot.

Each pupa is completely surrounded in a clear shell, which encloses the body tightly like a diving suit. This shell only really becomes apparent after it is shed by the young bee as it hatches (see also pages 46/47).

—

OPPOSITE
This can be observed time and again – bees remove the cover from the pupa cells, examine the pupae and seal up the cells again. The reasons for and purpose of this behaviour are unknown.

—

The combs of honey bees – marvels created as a collective effort of the bee colony. The combs serve as storage facilities for nectar, which the bees use to produce honey. During the production process, the bees fortify the nectar with an array of enzymes and remove water from it by heating it up. The resulting honey is then sealed inside the combs with wax caps. The bees mash pollen and mix it with the honey to create bee bread, an essential nutritional source for the bee colony. Combs also serve as nurseries; each larva grows up in its own cell. The pupa cells are sealed shut with a wax cap.

Three activities among many that can be seen: a young bee gnaws open the cap to its cell before hatching out of it in the 14th cell from the left and in the 10th row from the bottom. A cleaning bee scrubs an empty cell in the 13th cell from the right and the 10th row from the bottom. A worker bee stomps pollen tightly into the 10th cell from the right and the 4th row from the bottom.

—

ABOVE

Sealed cells in the brood nest. The flat caps give away that new worker bees are currently growing in the cells. The offspring in the sealed cells are supervised closely as well. Heater bees heat the brood nest continuously to roughly 35 degrees Centigrade (95 degrees Fahrenheit). During the summer, up to 2,000 young bees can hatch from their eggs in the brood nest on a single day and commence their work.

—

OPPOSITE

An uneven surface on the brood combs reveals that males, drones, are growing in the cells. In addition, the diameter of their cells is greater than that of the worker bee brood. The bee in the empty cell may be cleaning or sleeping – it is impossible to differentiate between laziness and diligence in the photo.

OVERLEAF

Two worker bees hatching. Remnants of the clear, skintight shell surrounding the bodies, legs, feelers and wings of the pupae cling to their heads. We can only speculate as to the purpose of this shell. Perhaps it provides protection against water loss? Bees that are ready to hatch begin gnawing open the lid to their 'isolation unit' from the inside. In order to open the lid completely, worker bees on the outside usually come to their aid.

—
LEFT

A fully-developed drone fights his way into the world. The feelers are already palpating the surroundings through the tiny opening and can immediately detect the smell of the nest. The eyes are of little use for orientation – it is pitch dark in the brood nest.

—
OPPOSITE

Hatching drone – remains of the pupa's burst body shell can be seen on one of the feelers. The eyes of the drones are enormous, the long feelers full of olfactory cells. A drone's sole objective is to serve as a living 'sperm bomb' by seizing a virgin queen that appears at a drone congregation area for mating and rapidly catapulting sperm into her in mid-flight. The senses of sight and smell work together to spot the queen high up in the air and grab her before the rest of the competition.

The drones are accepted during
the mating season when the
colonies are raising new queens.
Their muscular bodies can help
with thermoregulation, albeit as
an inadvertent side effect of their
presence. They are fed for a while in
order to keep up their strength.

Cleaning is an essential element in a honey bee's behavioural repertoire. Hygiene is at the top of the agenda in the bee colony. The drones are kept scrupulously clean, as otherwise they might bring dirt and pathogens back to the colony when returning to the hive from failed excursions in search of a queen. After all, the drones do not just fly around outside, they also sit around in the vegetation. When cleaning, parasites such as the Varroa mite are bitten and removed as well.

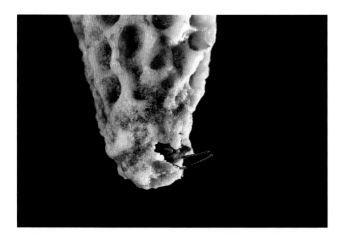

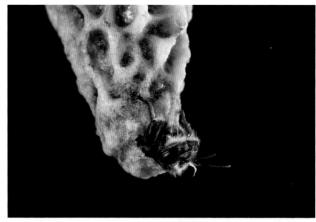

Queen   Drone   Worker

—

OPPOSITE

If a queen dies, a successor has to be found post-haste. For this purpose, a cell directly in the middle of a comb is elongated into what is known as a supersedure cell. However, this supersedure is only possible if there are still small larvae in the bee colony, since it is only possible at an early stage to change the path of a young larva into that of a queen.

—

ABOVE

A new queen sees the light of day. The workers have created a large, cone-shaped cell on the edge of a comb – a queen cradle. In contrast to the worker bee and drone larvae, the larvae that hatch there are fed exclusively royal jelly, the 'milk' of nurse bees produced in special glands.

SQUATTERS IN THE BEEHIVE

The dwellings in which humans keep honey bee colonies are called hives. Beekeepers generally see fellow lodgers in these hives as foes of the honey bee – first and foremost, the *Varroa* mite, a recently imported parasite. There are also the equally parasitic tracheal mite, the comb-eating lesser and greater wax moths and the small hive beetle, which devours pollen, honey and the brood, thus completely destroying the bees' nests. Beekeepers consider themselves responsible for protecting their bee colonies from these 'squatters' and do their utmost to combat them. Primarily, toxins are used for this purpose, and they give rise to a range of associated problems. The beekeepers aim to win a decisive victory over the parasites while preserving the bees, but because of the limited living space shared by these organisms, this is not an easy task. The toxins used should not end up in the wax, if at all possible, and definitely not in the honey. It therefore comes as no surprise that beekeepers are not fond of beehive squatters.

Are these, and maybe even other squatters, found in the nests of wild bee colonies? What types of flora and fauna grow in the hollow trees used by bee colonies to construct their combs? And what sort of interactions are there between the various life forms? These are questions which have only recently begun to be asked as interest in wild bee colonies and their living conditions is quite new.

Large cavities inside tree trunks, especially those left by woodpeckers with just a small opening to the outside world, are ideally suited to harbouring small, autonomous, virtually self-sustaining ecosystems. Thus, the bee colony socializes not only with microorganisms but also with multicellular organisms such as worms and articulated animals. In an ecosystem – even the microecosystem inside a tree cavity – the organisms living there are closely interlinked. And in contrast to artificial hives, the relationships between honey bees and their housemates are positive ones – bees in the forest have helpers.

Historical accounts from the times when bees were kept in straw baskets or log hives, i.e. in simple, hollowed-out pieces of tree trunk, and when the *Varroa* mite had not yet reached Europe, list the book scorpion as a commonly encountered housemate of bee colonies. It is a small arachnid measuring just a few millimetres in length, with pincers that give it the appearance of a true scorpion. Log hives and straw bee dwellings offer it a place to live

A book scorpion on a comb. From a zoological standpoint, the book scorpion belongs to the family of arachnids, which differ slightly from insects in that they have eight legs instead of six. Like insects, arachnids also belong to the phylum of arthropods.

ABOVE AND OPPOSITE

Book scorpions are frequently encountered as housemates in forest bee nests. They have also been observed clinging to bees departing from the hive in a swarm, thus accompanying the bees right from the start as they colonize a new nest. In the early years of beekeeping, this animal was also commonly found in straw hives, and it was held in high regard as a predator of bee parasites. If the right living conditions are created for it in the beehive, it can serve as an aide to the beekeeper, even against the Varroa mite.

A Varroa *mite walking across the abdomen of a worker bee. This parasite, which was transferred to the western honey bee from its original host the closely related Asian honey bee, represents the greatest threat to our bee colonies.*

Mutual cleaning is among the most important activities in the bee colony. In the close confines of the nest, hygiene is an absolute must. In the process, parasites such as the Varroa mite are eliminated. Bee breeders are striving to make genetic modifications in bees to achieve a more pronounced cleaning behaviour.

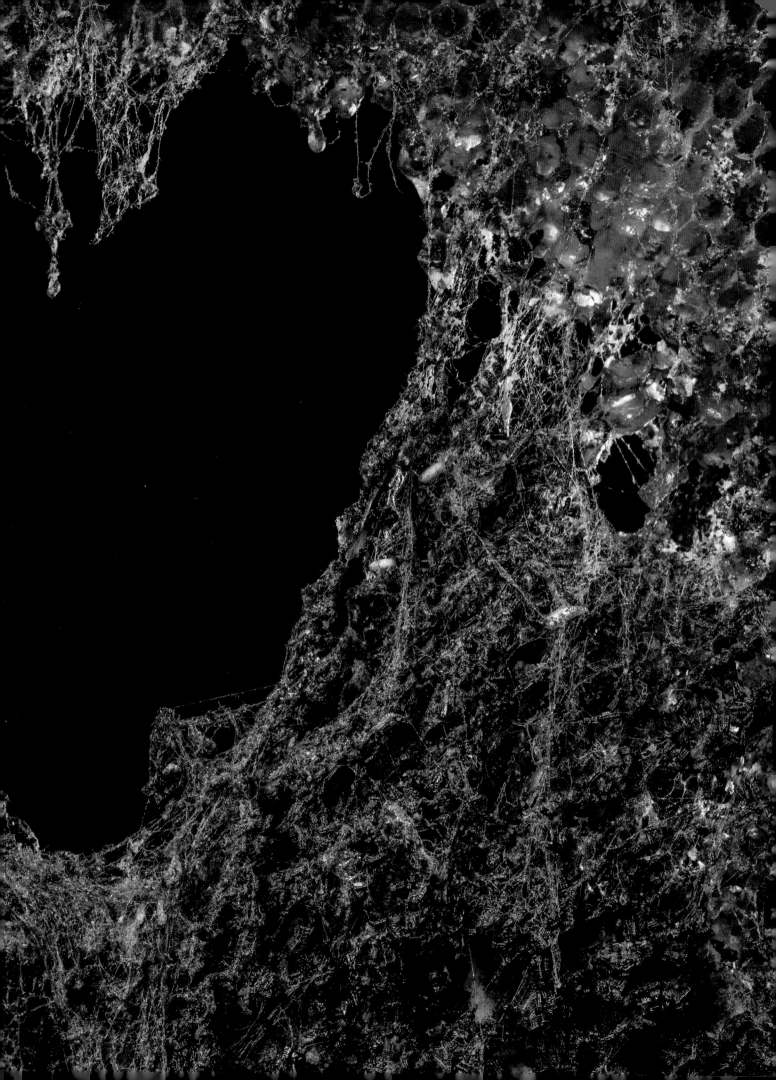

within the living environment of the honey bees. Here, it is able to hide, propagate and hunt for even tinier prey. Beekeepers were happy to see it. They treasured it as an annihilator of the tracheal mite, a small parasite that lives in the respiratory tract, the tracheae, of the bees and is on the book scorpion's menu.

Modern beehives do not however provide appropriate living space for book scorpions. These geometrically exact structures, made of planed wooden boards, do not feature any crevices or cracks to which they could retreat, and do not offer any sustenance for them during their juvenile stage, when they live off even tinier prey such as book lice. And book scorpions certainly cannot survive the chemical methods in use today to combat the *Varroa* mite. They perish long before the quite robust *Varroa* mite is eliminated. Book scorpions have disappeared from our beehives. There are, however, some exceptions. It is not uncommon for beekeepers, who set up beehives in the forest and are not able to look after them for a while, to find specimens of these little pincer-bearing creatures there.

In contrast, a perusal of the floor inside a tree cavity inhabited by bees, i.e. everything that has sprinkled down from above, frequently reveals book scorpions. They appear to be a regular housemate of the colonies of wild bees. Indeed, there are practically no tree cavities artificially created by a Zeidler, a traditional honey hunter from Germany (see

chapter 'Traditional Beekeeping'), that are not home to the book scorpion. It is logical to assume that they go hunting inside the tree hollows, where they come across prey whose eradication is beneficial to the bees. One need only think about the observations from previous generations of beekeepers on the tracheal mite. If pseudoscorpions are introduced into a *Varroa*-infested bee colony, the population of the parasites plunges rapidly. In artificial, experimental situations, individual book scorpions have slain up to ten *Varroa* mites per day. The mobile finger on each of the two pincers features a venom gland, which kills, or at least paralyzes, the mites. The pseudoscorpion then sucks down its prey, in true spider fashion. The book scorpion is a true friend of the honey bee.

Beekeepers dread the appearance of two species of moths, the lesser and greater wax moth. Their caterpillars feed on wax and on the membrane left behind by honey bee larvae. The caterpillars eat their way straight through the honeycomb, constructing tunnels lined with their own silk as they burrow. The result is a completely destroyed honeycomb. The creatures leave behind a chaotic mass of material, held together by the threads spun from their silk glands. A disaster for beekeepers.

But what happens in the forest? If a bee colony dies, the old honeycomb lingers on. The house is still 'furnished', full of old fixtures which nobody wants to look after, and which stand squarely in the way of a new bee swarm moving in. Furthermore, the combs which were used by the bees for many years are often infested with various pathogens of assorted bee diseases. Such a tree cavity with abandoned combs from many years ago is not an option as a home for a new bee swarm, and should no longer be used by honey bees either for hygienic reasons. But housing is scarce in the forest and suitable tree cavities are a precious resource.

This is where the wax moths come into play. The smell of wax lures the moths as if by magic. The females lay several hundred eggs on the combs and the resulting caterpillars eliminate the old combs completely, freeing up space for a fresh bee colony to start anew. In the forest, the pests so feared by the beekeepers act as a demolition crew, clearing the way for new residents.

So these studies of bee colonies in their natural environment reveal that the presence and actions of housemates, which are considered a curse by beekeepers keeping bee colonies in beehives, are actually a blessing for bee colonies living in the forest.

Very little is known about the composition of the bacterial and fungal communities and their association with honey bees in hollow trees. We suspect that tree hollows provide an ideal environment for harmful bacteria and fungi to grow and thrive, and that this leads to considerable hygienic efforts by the bees. These include the use of anti-fungal and anti-bacterial propolis, which the bees produce as a resinous secretion from tree buds. However, we have absolutely no idea whether any of these microorganisms help the bees in this regard or not.

Can the study of the mini-ecosystem formed by bee nests teach us something for practical beekeeping and can we adopt individual aspects in the field of apiculture in order to make the lives of bees and beekeepers alike easier? The book scorpion could serve as a starting point.

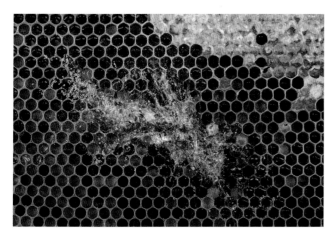

—

TOP LEFT AND TOP RIGHT

The greater wax moth (left) and the lesser wax moth (right) are considered pests and handled as such by beekeepers due to the destruction wreaked by their larvae on the combs.

—

BOTTOM LEFT AND BOTTOM RIGHT

In nature, greater and lesser wax moths are important partners for the honey bees. By destroying the old, vacated nests, the moth larvae create room for a new comb to be constructed.

TOP LEFT

Certain housemates help the bee colony to channel dead members, that cannot be removed from the tree cavity, back into the material life cycle. Skin beetles and their larvae are included here as scavengers. They feed on bee carcasses and their resulting faeces contribute to creating a layer of sediment, which in turn supports other organisms.

—

MIDDLE

Tree cavities are ideal retreats for small creatures, which either reside there every once in a while, like this spider, or which take up permanent residence in the layer of sediment at the bottom of the cavity, such as woodlice or small worms.

—

BOTTOM LEFT

Ants and honey bees represent important members of the ecological network in the forest. Aphid colonies supply honeydew and are both guarded and cared for by the ants. Ants can be encountered regularly in the nest cavities. Maybe they are also beneficial to the bees? We have no idea..

A handful of sediment from a tree
cavity inhabited by a bee colony
continuously for the last three years.
It is like the bottom of the deep
sea - everything that trickles down
from the bee colony is utilized by
the organisms living in the sediment.
Leftover wax, dead bees, pollen and
fragments of propolis find grateful
users here, which are dependent
on this 'rain' from above. The mini
ecosystem created as a result of
this provides biological balance,
which also benefits the bees. No
decomposition or decay processes
take place, which would otherwise be
a source of disease.

DEFENCE AT ANY PRICE

Living in hollow trees not only protects honey bees from rain, storms, cold and heat. It also protects them from enemies. A bee nest, from an objective standpoint, is a highly concentrated storage facility full of valuable resources: easily digestible carbohydrates in the form of honey, and, in the plump larvae and pupae, a rich source of protein. And the adult bees themselves represent an attractive bounty for predators in terms of their sheer numbers and their extremely high density.

The relationships between predators and their prey are some of the most powerful drivers for the evolution of living things. In some cases, over the course of time, strategies and counterstrategies produce extraordinary results through natural selection on both sides. In this 'arms race', however, the pressure to succeed has very different direct consequences. A failed attempt merely means a missed meal for a predator; things can go better during the next assault. For the prey on the other hand, the predator's success almost certainly means its demise – there is no second chance!

As an omnivore, a bear in search of honey has many alternatives to the sweet snack should it fail to clear out a bee nest it has discovered. A bee colony that falls prey to a bear however is sure to perish. Over very long periods of time living together in the forest, this 'beary' threat has led to the bees developing the ability to recognize bears based on just a few key features. Dark colouring, a fluffy body and breath rich in carbon dioxide betray the advancing bear long before it begins climbing up the tree. The bee colony has been warned and tens of thousands of stingers await the bear's arrival. Not every bear is frightened off by the scores of bee stingers, but not every bear attack is victorious either. The defenders have a good chance of success.

Bright clothing, smooth fabric and not breathing directly into the open beehive are beneficial rules of thumb followed by beekeepers when dealing with their bees, and their effectiveness can be traced back to the bee–bear relationship.

If bees have a choice, they always pick dwellings that are several metres above the ground, which is surely a safeguard against large predators such as brown bears. The far-field vision of bears is rather poor, so they often fail to spy the nest entrances of honey bee colonies when they are located high in trees.

—

Two guard bees standing vigilantly. Their mouthparts, the mandibles, are spread and ready to bite, the fronts of their bodies are raised menacingly. Worker bees prefer to linger around the entrance to the nest between their activities inside the hive and in the field and are therefore automatically the first bees to encounter foes.

Outside the nest, honey bees are at the mercy of wasps and hornets, their natural enemies. Once a bee has been captured, it is often stung with precision between its body segments. The wings, feelers and legs are then snapped off, while the head, thorax and abdomen are flown back to the wasp nest in pieces..

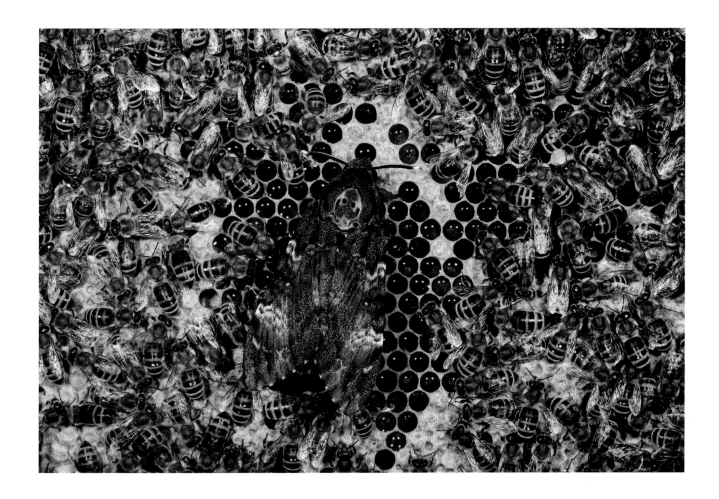

A death's-head hawk-moth – a goliath undisturbed amongst dwarfs. This is baffling, since death's-head hawk-moths feed on the honey from bee combs. They pierce the caps on the honey cells with their strong proboscis and help themselves to the contents. The intruder weighs as much as 200 bees. But it is by no means its size that keeps the bees at bay. Any mice that infiltrate the hive are stung to death by the bees, and they are a great deal larger than the death's-head hawk-moth. A chemical cape of invisibility enveloping the moth is what saves it, selling itself as an oversized nest mate.

»IF FOOD BECOMES SCARCE IN LATE SUMMER, HONEY BEES WILL TURN INTO BANDITS, ROBBING THEIR FELLOW BEES.«

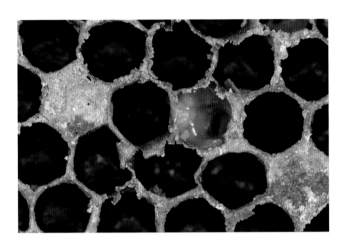

—

OPPOSITE

The worst enemies of the bees are the bees themselves, not unlike humans. If the provisions in a colony are running low and the supply of nectar is sparse, robberies are likely to occur. Bee colonies raid one another. Here, the workers from a robber colony have massacred the workers of another colony and are now loading up on the dead colony's honey.

—

ABOVE LEFT AND RIGHT

The marauding bees work rigorously, undisturbed. They killed all the defenders during the raid and are now loading up on honey. Not every raid is a success. If the bandits are outnumbered by the target, the defending colony will strike back.

But a house high up in a tree trunk does little to protect them from winged adversaries. The worst enemies of the bees in this aerial brigade are actually their close kin – wasps, and especially hornets, which actively hunt adult honey bees. These predatory insects, belonging to the order Hymenoptera just like the honey bee, are primarily carnivores. The fully-grown hornets or wasps and their larvae have a mutually dependent relationship when it comes to feeding. The 'wasp waist' of adult hornets and wasps prevents large pieces of prey from reaching the gut where digestion takes place. And so the hornets and wasps have come up with a complicated, yet ingenious solution to this problem. Prey such as the honey bee is torn apart, after which the chunks are flown back to the nest and fed to the larvae. The larvae predigest the pieces, regurgitate the resulting juice and then feed it to their grown sisters.

Away from the nest, the honey bees are at a hopeless disadvantage to the hornets and wasps. The slick abdomens of the latter offer only a small area for the bees to attack, while their highly flexible bodies allow them to sting the bees from between their own front legs, with their sharp-edged mouthparts, the mandibles, working like surgical instruments.

If a hornet intrudes into a bee nest, however, the cards are redealt. Now the circumstances are reversed. If the guard bees standing at the entrance to the nest sound the alarm, the whole scenario can play out directly in front of the nest, taking place inside at the very last minute. Such intruders are cooked, in the truest sense of the word. The bandits are not seized, stung or bitten. They are simply wrapped tightly in bee bodies. This ball, with the enemy tucked away in the middle, then starts to heat up. The bees manage this using the same method employed by the brood nest heater bees to thermoregulate the pupae and by the nectar-thickening bees to create honey: they generate heat by vibrating their strong flight muscles. In doing so, the temperature can reach up to 44 degrees Centigrade (111.2 degrees Fahrenheit), as measured on the body surface of the bees. If it were any higher, the bees themselves would not survive. For the hornets and wasps enveloped in heating bees, such a high temperature is too much to bear. They cannot endure this 'hug'. The enemy is killed by the community without a single stinger having to be used. The narrow temperature window between 'death for hornets/wasps' and 'death for bees' is the bees' saving grace. The body proteins of the victims solidify at the high temperatures. The fact that no oxygen can reach the engulfed trespasser helps to guarantee that it cannot escape. Once the defensive manoeuvre has been successfully completed, the ball of bees disperses and the dead hornets are thrown out of the nest.

However, there are also considerably larger intruders that can move around inside the bee colony unchecked using a cape of invisibility. The death's-head hawk-moth is one

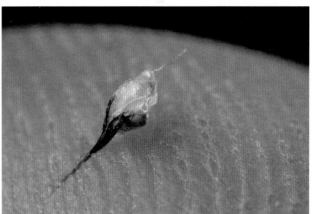

—
ABOVE

The honey bee's stinger evolved from the ovipositor of its predecessors millions of years ago. That is why drones do not have a stinger. The stinger, usually concealed inside the bee's body, is composed of multiple parts, in contrast to the needle of a syringe.

—
RIGHT

Once a bee has delivered a sting, its stinger is ripped out of its abdomen as it flees. It does not survive this. The venom sac is still hanging on the stinger. Tiny muscles allow the two halves of the stinger to push against one another, its jagged edges drawing it deeper and deeper into the victim, like barbs.

—

OPPOSITE

Hornets know exactly where bees are an easy target. They persistently patrol stockpiles of flowers, lying in wait for careless bees. And even attentive bees are hardly able to escape the hunters in flight.

—

ABOVE

Once a hornet has a bee in its clutches with its strong mandibles, the bee is doomed. The hornet's highly manoeuvrable abdomen allows it to sting in all directions, even to the front. If hornets discover a bee nest, an outright slaughter commences, against which the bees are completely helpless.

of the largest moths in Central Europe. Its menacing name comes from the striking, white, vaguely skull-shaped pattern adorning its dark thorax. Swedish botanist Carl Linnaeus gave the hawk-moth its scientific name *Sphinx atropos* during the eighteenth century, inspired by this mark – in Greek mythology, the goddess of fate, Atropos, cuts the thread of life at the hour of death.

The caterpillars of the death's-head hawk-moth feed on nightshades together with a wide range of shrubs, such as privets, lilacs, oleanders and datura. For the growing hawk-moths, the honey bee nests represent the most important source of sustenance. There, they live off nectar and honey. If you are lucky enough to see one of these giants sneaking across the combs between the bees undetected, you might be dumbfounded by the fact that the bees tolerate its presence, allowing it to puncture the caps of the sealed honey storage cells with its strong proboscis and help itself to the contents. It makes quick work of it all. In three minutes, a death's-head hawk-moth can drain an entire honey cell dry.

Hawk-moths are capable of generating loud squeaking sounds that are even audible to the human ear. This type of sound production is unique among the insects. They do this by sucking air into their foregut and creating loud sounds when 'inhaling'. A second, softer sound is generated when the air is let out of the foregut through the mouth.

Many a beekeeper has heard this concert. When it was
observed that these intruders are left alone by the bees,
it was first conjectured that the sounds emitted by the
intruder keep the bees at bay, since even a slight brush
against its body is enough to trigger the squeaking. In fact,
however, the death's-head hawk-moth has a chemical cape
of invisibility. The surface of its body is covered in a mixture
of fatty acids whose composition, in terms of mixing ratio
and concentration, is virtually identical to the mixture taht
naturally coats honey bees. The perfect camouflage. In the
dark beehive, the bees cannot recognize who is jostling by.

Although the guard bees at the hive entrance are aware
that a giant is breaking into the hive, stings from a small
group of bees do not pose a real threat to the death's-head
hawk-moth, which can survive up to four bee stings at
once, unscathed. And once inside the nest, out and about
amidst the combs, it is not in danger of being cooked alive
like a hornet because, as far as the bees in the nest are
concerned, it does not even exist.

—
OVERLEAF
*Guard bees at the entrance to the
nest have spotted a hornet. If an
enemy tries to infiltrate the nest,
the bees will attempt to ward it off
together. Foes that have broken
into the tree cavity are attacked
from inside the nest, with the bees
surrounding them by stretching
their bodies across the combs and
interlocking their legs.*

A hornet, an invincible mortal enemy
of the honey bee outside the nest, falls
victim to the concerted defence effort
of many honey bees at the entrance
of the bee nest.

The bees immediately recognize the intruding hornet as an enemy, but they do not attack it with bites and stings. The bees have another means of defence at the ready.

The hornet is quickly wrapped in bee
bodies, which crowd tightly around
the foe in multiple layers. All ongoing
activities are interrupted for these
'hugs'.

The bees raise their own body temperature until just below the limit that would be fatal to them, but above the temperature at which hornets and wasps can survive.

The hornet's desperate attempts
to escape the tight embrace by
biting and stinging are futile. It dies
of hyperthermia and asphyxiation,
buried below countless bees, within
just a few minutes.

Once the hornet is dead, the ball of
heat dissipates, and it is dragged
out of the beehive or dropped onto
the floor of the cavity, where the
honey bee's housemates take care of
degrading the corpse.

THE FOREST HABITAT

The appearance of the honey bee's forest habitat is governed primarily by the climate and humans. Before human intervention, forests formed over very long periods of time influenced by alternating glacial periods and interglacial periods. The climate determined which tree species were most prolific and how they were distributed geographically. As our ancestors developed settled communities, humans began shaping and transforming the forests. Their use, first, as forest pastures and then, later, as commercial forests to produce timber has led to the forests we know today. And the honey bee has been ever-present.

The forebears of the current western honey bees, which all belong to the species *Apis mellifera*, migrated at least 25 million years ago from Africa to Europe and spread rapidly. However, a stable bee population was never established as the glacial periods pushed the bees time after time back towards southern Europe, from where they progressed again north as the ice sheets retreated.

Inevitably, the ability to generate heat helps combat a colder climate. Flying insects such as honey bees possess a perfect heat generator in the form of their flying muscles, jam-packed into the body section bearing their wings and legs – the thorax. This heat source is used by all species of honey bees to warm the pupae in the brood nest, even those that live in sub-tropical and tropical regions. This demonstrates an ingenious trick of nature

– once an 'ability' has been acquired, it is often then used in completely different ways in addition to the benefit it first brought. For example, with honey bees, the ability to generate warmth is used as 'heating' during the cold winter months. This collectively operated heat production is what allows bee colonies to survive all of the seasons – in contrast to wasps or bumble bees, whose entire infantry dies off before the onset of winter, and only the young, fertilized queens await the spring, hidden in hibernation and protected from frost.

If honey bees were to build their combs outdoors in termperate latitudes, their energy reserves in the form of honey would not be sufficient to deal with the numerous weeks of below freezing temperatures. Quite simply, the bees would not survive the winter.

This is where the forest comes into play. Hollow trees offer an ideal living environment for honey bees. They offer them protection from predators such as bears, for which the honey represents an irresistible target. If bees are able to choose between artificial nest boxes hung at different heights, they tend to prefer one located 6

Blooming bluebells carpet the floor of a beech forest – a land of milk and honey for bees.

metres (20 feet) above the ground. This height poses a huge challenge for bears in search of honey, thus granting the bee colony a degree of relative security. Furthermore, cavities inside trees can have a world of their own. A climate is established inside which can then be shaped and stabilized by the bees, and there is also an ecosystem connecting the bees to a myriad of other living creatures within a social construct.

A healthy forest provides a constant supply of plant products from spring to autumn, which a bee colony needs in order to survive and propagate. This is unlike the open countryside, where the availability of nectar, meaning all the provisions the bees can bring back to the hive, is becoming increasingly one-sided, if it is even available at all. One of the unpleasant consequences of this development is that bee colonies can starve to death in the summer.

What extra does the forest have to offer the bees? Probably the best-known term for describing the relationship between bees and the forest is the one that is the most succinct –'forest honey'. It may come as some surprise to hear that forest honey does not come from flowers. In fact, it is based on excretions from aphids and scale insects – which is actually not as disgusting as it first sounds. Upon closer observation, the origins of forest honey open our eyes to an extremely fascinating triangular relationship between ants, aphids and honey bees.

Trees absorb water and salt from the soil, which are then transported through minuscule pipelines to their highest leaves. There, through the process of photosynthesis, the components mix with carbon dioxide absorbed from the atmosphere to create glucose. The glucose solution produced in the process wanders just below the bark in sieve tubes to places in the plant that require the sugar as a raw material to be converted into cellulose, starch and much more. Small beneficiaries target these highly valuable miniature flows. Aphids tap into the sieve tubes by sinking their mouthparts, which have evolved into proboscises, directly into these tiny pipelines. However, they are not really after the sugar but rather the amino acids, nitrogen-rich compounds that they need to metabolize proteins. But what do they do with all the sugar? The aphid's body is designed to excrete the absorbed, but unnecessary, material as a clear sugary solution through special excretory glands on the abdomen.

This substance is known as honeydew and is very popular with the social insects of the forest. The relationship between the aphids and the ants is the closest. The latter stroke the abdomen of the aphids with their feelers, and the aphids then excrete the honeydew and are thus milked by the ants. This source of food is so important for ant colonies that they actually go to the trouble of collecting aphids; they frighten away predators such as other types of insects, spiders or even birds.

The honey bees in turn profit from this symbiosis. The honeydew can form thick, sticky coatings on needles and leaves, which the bees collect and transport back to the nest in their honey stomach. Even though there is a variety of specialist aphid species found on conifers and deciduous trees, honeydew is still not as readily available over time and space as nectar.

Ferns existed in forests long before the appearance of the first flowering plants. And heavily laden honey bees flying slowly away from a fern frond having collected nectar in its original form are not an uncommon site. Certain glandular areas of tissue, which can be dispersed across the entire plant, secrete nectar. The ferns clearly do not do this to attract pollinating insects, since ferns do not have flowers in need of pollination. In the case of more evolved flowering plants, which depend on pollinating insects, nectar-producing

An untrained eye will scarcely notice the honeydew-dispensing scale insects gathered in thick, brown clumps between the pine needles. However, they do not remain hidden from questing honey bees, making an important contribution to the production of honey in the forest.

>»NATURAL, HEALTHY FORESTS BOAST
A HIGH LEVEL OF BIODIVERSITY IN TERMS OF
FLORA AND FAUNA. THIS INCLUDES FLOWERING
PLANTS AND THEIR POLLINATORS.«

cells, or nectaries, are concentrated on their flowers. There, it is in their interest that the bees visit the flower.

How does the forest benefit from the bees? Natural, healthy forests boast a high level of biodiversity in terms of flora and fauna. Flowering plants, which rely on pollinating insects, represent a significant portion of this biodiversity. If important insect species disappear, honey bees can take over their chores as well. In this case, it is helpful that they have a flying range 10 to 20 times larger than that of a solitary forest bee, extending many kilometres. Furthermore, they are not limited to specific flowers, being able to visit virtually any of them.

Pollinated flowers produce fruit and seeds, which are used to sustain many bird species in particular. But even the bees themselves offer a relevant source of food. Throughout the course of a year, a bee colony spawns roughly 20 kilogrammes (44 pounds) of biomass in the form of all its members, of which the majority die in the vicinity of the bee nest and then serve as food for ants, wasps, birds and other animals.

Honey bee activities also include sanitary tasks, collecting honeydew prevents black fungus and mildew from forming on the leaves and needles of the trees, as the coat of honeydew would otherwise offer the fungi a welcoming breeding ground. And even bee colonies that have disappeared entirely can be good for the forest. The abandoned, empty combs are colonized by wax moths, whose caterpillars in turn can be used to nourish the ant beetle, a nemesis of the rightly feared bark beetle.

OPPOSITE AND ABOVE
Linden trees are in full blossom during the summer. None of the countless flowers is left unvisited. Linden flowers supply the bees with large quantities of nectar, thus allowing them to increase their reserves of food considerably within just a short period of time.

—
CLOCKWISE
Corydalis in a forest clearing, honey bee on groundsel, wild strawberry flower.

—

—

CLOCKWISE

Forget-me-not, blackberry flower, honey bee on a willow catkin, wood anemone, foxglove.

—

OVERLEAF

Among the flowering woodland plants, the blackberry is one of the most important sources of food for the honey bee.

—

Forest fires are one of the greatest threats to wild honey bee colonies. The presence of smoke acts as a signal to fill up with honey, retreat deep into the nest between the comb alleys and wait. Beekeepers make use of this reaction when they pump smoke into an open beehive with a smoker (see page 181). During a forest fire, fleeing makes no sense for the bees since the queen gains tremendous weight after her nuptial flight and is no longer able to fly until she leaves the nest in a swarm. If the colony remains steadfastly in the nest and surrounds the queen in thick clumps, the odds are not bad that the queen and many of the worker bees will survive the exposure to the outside heat. One can only speculate as to the reason behind filling up on excessive amounts of honey. They may be exploiting a physical property of matter – a bee body brimming with honey has a higher heat capacity, allowing it to absorb a greater amount of heat with a lower increase in body temperature, thus delaying or even preventing the fatal effect of the heat.

»PROPOLIS, COLLECTED BY THE BEES
ON BUDS, HAS ANTI-BACTERIAL AND
ANTI-FUNGAL PROPERTIES, AND SERVES
AS FIRE PROTECTION.«

—
TOP LEFT
A bee collecting propolis before lifting off to fly home.

—
ABOVE
Like the buds of many other plants, the leaf buds of chestnuts provide the basic element for making the valuable, anti-bacterial and anti-fungal propolis – here the resinous plant secretion is clearly visible as a shiny coating on the bud.

—
ABOVE LEFT
Once the interior walls of the newly colonized tree cavity have been smoothed out sufficiently, they are lined with propolis by the bees.

—
OPPOSITE
If a forest fire moves on rapidly, so that only the surface of the trees is burnt, the bees will survive it.

—

OPPOSITE

The tree cavity provides optimal conditions for the bee colony, even during winter. The thick walls and small entrance hole keep the cold outside, which means the bees only have to burn a small amount of honey in order to keep from cooling down to below 10 degrees Centigrade (50 degrees Fahrenheit). This allows them to remain mobile.

—

ABOVE

Warm periods in the middle of winter are fatal for honey bees. The bees dissolve the winter cluster, in which they were tightly packed together, and the first forager bees leave the nest. If they stay outside for too long, and the weather worsens, they will become stiff and unable to fly. They then perish on the snowy surfaces.

ORIENTATION WITH ALL THE SENSES

The forest is a natural environment in which we humans can easily get lost. Without paths, markers and a clear view of the position of the sun (and today without a map or GPS), pinpointing locations or finding the desired direction is extremely difficult, if not impossible. Hansel and Gretel faced this exact problem in the well-known fairytale of the same name and laid down a trail of white pebbles to help them find their way back. During their second sortie into the forest, they foolishly used breadcrumbs instead of pebbles and the birds ate the crumbs. They got lost in the woods.

For beetles, butterflies and other insects that do not have a permanent address to find, it is irrelevant that the forest is difficult to navigate. Ants on the other hand, which have to find their way back to their nest after outings, use endogenous glands to leave scent trails on the ground for them to follow and so avoid getting lost.

But how do honey bees navigate in the forest? We are not exactly sure. We only know that it works perfectly as they do not lose their way. Until now, the orientation abilities of honey bees have been studied under conditions where the researcher was able to maintain a clear overview – in open spaces, on meadows, even on decommissioned airfields in some cases. If a forest was included in the study, then the observations took place on its edges or along straight forest paths, but never in the depths of the forest itself. The prevailing opinions on how honey bees find their way

home assert that they orientate themselves on the position of the sun and on polarization patterns in the sky, which are invisible to the human eye, as well as by remembering landmarks in the vicinity of the beehive. When visiting more distant destinations, a veritable map forms in the bee's head and this can then be conveyed to nest-mates using the dance language, through which other bees can interpret the precise coordinates of the target based on the dance, thus allowing them to find the location. This however is a hypothesis that lacks convincing support. The scents of the flowers near the target (an unresolved issue in all previous studies is what is considered 'near') and the smell of the beehive, including pheromones, also serve to help the bees in orientation.

But do these guides work in the forest? To date we know virtually nothing about whether they do. This is why an

—

A forager bee has discovered a new source of food. By performing a waggle dance, she draws attention to herself and conveys the rough location. Intrigued followers then take off in the indicated direction where subsequent stimuli such as alluring scents and ostentatious flying patterns, combined with continued communication, lead them to the target.

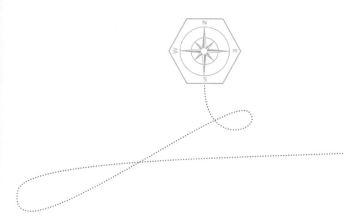

investigation into the topic would be extremely interesting and important since the forest, as the ancestral home of the honey bee, provides the conditions under which the orientation skills and communication abilities of these animals developed.

Let us attempt an approximation. A glimpse into the forest below the treetops generally offers us humans very few striking landmarks (one is unable 'to see the forest for the trees'). A look up into the canopy does little to help us find our way either. Similar perspectives are afforded to the flying honey bees through their sense of sight as they whiz through the trees close to their goals. Of course bees also have the benefit of a view from above onto the sea of treetops, but that does not appear any less confusing.

However, there is a definite difference between human perception and bee perception. The latter are capable of seeing polarized light, which is light that oscillates in a single plane. In principle, sunlight oscillates in all directions perpendicular to the direction of propagation, but it is scattered and reflected by the atmosphere, thus exhibiting preferred directions of oscillation across the sky. Because bees are able to perceive this, they can discern the position of the sun and thus orientate themselves even under a partly cloudy sky where small patches of open sky are visible. The polarization of the light produces a pattern in the sky which changes little by little from one place to another, although neighbouring regions are hardly distinguishable from one another. This could be easier for bees living in the forest. Because the sunlight is dispersed into small fields under the canopy of leaves, the directions of vibration could be more clearly visible to the bees. However, no studies have broached this subject to date.

How do scents assist in orientation and navigation inside the forest? There is a dearth of studies and data on this as well. Behavioural observations of honey bees and plausibility considerations have provided initial indications. Scents act as highly suitable guides when an air movement takes them from the destination to the bee. The bee merely has to fly against the wind. However, if the wind is blowing away from the bee, a scented target is of no use.

But the forest has its own special laws in that respect. Air movements in the forest, wind in other words, are buffered considerably, allowing the forest to offer a windless space. It is possible to experience and smell how scent trails behave in the windless forest. If you find yourself on a path along which a smoker was strolling earlier, even a long time ago, the odour sticks to the vegetation in clouds for a long time thereafter. Pheromone traps used in the forest to fight bark beetles by attracting the male specimens work across great distances.

When the worker bees discover new, fertile sources of food (which are often scarce within the spatial limits of

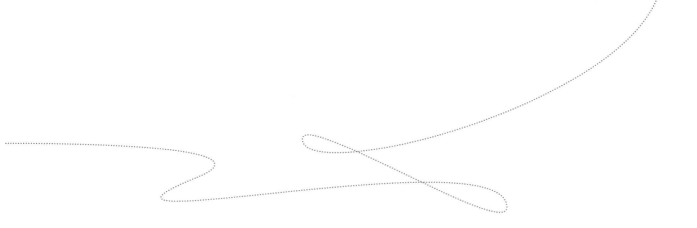

the forest) such as blossoming trees, blackberry bushes or colonies of aphids dispensing honeydew, they recruit helpers to accompany them to that target. A form of communication used by the bees is the dance language. Despite all of its inaccuracies and shortcomings, this is the key way by which the experienced bees, which know the target, can communicate its whereabouts to the recruits, which are eager to reach it. As long ago as 1923, the old master of research on bee behaviour, Karl von Frisch, described how the same forager bees that he had observed dancing in the observation hives could be seen performing striking flagging flights around the target while keeping a gland open on their abdomens. Geraniol is discharged from this gland (known as the Nasanov gland), which is an irresistible lure for the bees.

Would it not be conceivable that the experienced bee leaves geraniol in the air, not only close to the destination, where it can be easily picked up, but also long before reaching the destination, thus leaving scented markers which lead other bees to the target? Yet another gap in our knowledge.

If you observe the behaviour of experienced bees and recruits arriving at flowers or at an artificial feeding station in the forest, you will discover pairs of bees flying in tandem, with the bee in tow often landing on the lead bee and not on the flower. If you train forager bees to visit an artificial feeding station in the forest, then you

will also, after a while, find recruits at that location. If the light conditions make it possible to see the incoming bees flying low between the trees at a distance of a few metres, it becomes evident that their trajectories are not evenly distributed to the right and left of certain tree trunks, but rather most of the bees follow the same routes to reach the feeding station. Recruits tend to follow closely behind experienced bees. This represents a fascinating challenge for the scientific community – do the recruits follow the scent trails of the experienced bees, stuck firmly on the windless vegetation?

Every beekeeper and forest wayfarer is familiar with a humming, buzzing forest. While the honeydew produced by an aphid colony is being harvested by honey bees high above in the treetops, one can hear the flying sounds of the bees radiating like a noisy roar. Behavioural experiments have revealed that experienced foragers use these so-called flagging flights to attract recruits, in particular when the target has a weak scent or none at all. Aphids are not known for having a noticeable scent of their own. Do experienced bees help identify their location by emitting clouds of geraniol? Future studies will have to show whether the 'roaring forest' can be attributed to this behaviour.

The interplay between communicating through dance and marking targets is well known within another context in the life of bees – swarming.

A glance up into the canopy of a forest paints an entirely different picture of the sky than a look from an open field (see page 112). The sky here seems to be a pattern of spots. It is conceivable that this speckled sky makes it easier for the bees to determine the direction of the sun than the view of a sky in which neighbouring points have a virtually identical optical appearance. If the sun is covered by clouds, or by treetops in the forest, the bee's ability to see polarized light helps guide the way. Polarized light is produced as a result of sunlight being scattered in the atmosphere, creating a pattern across the entire sky and forming a ring around the sun. This makes it possible to determine the position of the sun from anywhere, as it lies perpendicular to the direction of oscillation of the aligned polarized light waves. The shorter the wavelengths in the light spectrum the more stable the polarization pattern will be against disturbances caused by drops of water and other small airborne particles. Thus, it is understandable why the bees would use the shortest wavelengths to orient themselves based on cues in the sky, which they can barely discern and which are just invisible ultraviolet light for us humans.

Honey bees use their sense of sight
to orient their flights with the sun
as a point of reference, using it as
a compass. An internal clock, which
takes into account the movement of
the sun, helps the bees to get their
bearings even if prolonged periods of
time have elapsed between several
flights to the same target.

Blooming linden trees rule this forest in which bees need to find their way. Whether soaring above the treetops or zigzagging through the trees, it becomes immediately apparent that finding a destination here is no easy task. Evolution has gifted the honey bee with all the characteristics needed to solve this conundrum.

ABOVE

ABOVE

Experienced forager bees fly back and forth between a rich source of food and their nest. In order to recruit helpers, they perform dances on the combs (see page 107), thus encouraging intrigued bees to fly into the area they have discovered. The recruits are provided with assistance in finding the destination in other ways as well. Recruited bees have even been seen appearing to latch on to an experienced comrade and allowing themselves to be guided to the feeding site by flying in tandem.

—

OVERLEAF

A forager bee bingeing on a flower loaded with pollen.

OPPOSITE

Two forager bees on the way to a food source. The wide-open Nasanov gland, from which they emit the pheromone geraniol while flying, is clearly visible. In this manner, they mark the trail to the feeding site – a valuable complement to the waggle dance.

Bees even mark the entrance to their nest with the pheromone geraniol, which they discharge backwards by whirring their wings in a standing position. This makes it easier for their nest mates to find their way back to the nest.

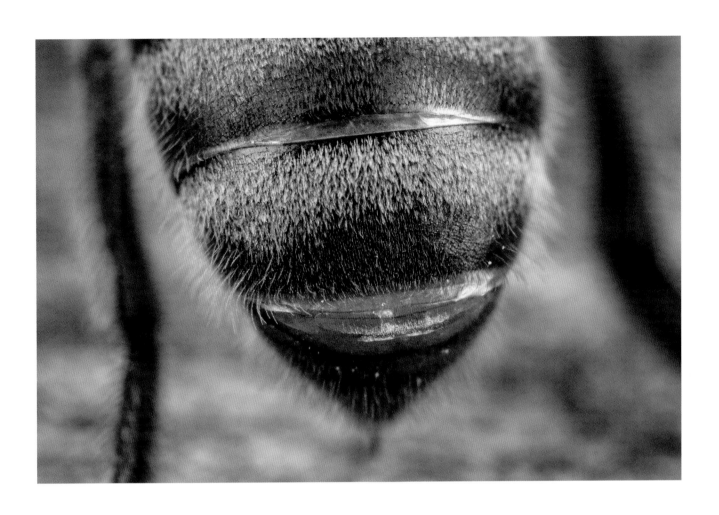

The Nasanov gland, located between
the penultimate and last body
segment on the abdomen of forager
bees, can be seen with the naked eye
when completely open.

—

ABOVE

A long exposure time reveals that a swarming cluster consists of extremely active bees that are just waiting in anticipation to find out when they can set off to a new dwelling.

—

OPPOSITE

An imposing swarm of bees has settled down not far from the old nest. From this temporary camp, bee scouts search for the best new abode available.

Bee researcher Patrick Kohl films the dances performed by successful bee scouts on the surface of the swarm cluster in Hainich National Park, Thuringia, Germany. The objective of this study of a swarm, settled on an artificial base, is to find out how many suitable new nesting sites have been discovered by the bee scouts within a distance of several kilometres.
In addition, their advertising dances provide the researcher with indications as to where these nest cavities might be located. However, the swarm does not find the new nesting site based on the dances, most of the bees are not even aware of them. The bees that know the location lead the rest of the swarm. Furthermore, they have marked the target with pheromones. Because the queen was placed in a small cage by the investigator directly in the centre of the swarm, these test bees are sure to return again after they have left. Without the queen, they would not be able to form a new bee colony.

»SOPHISTICATED EXPERIMENTS HELP US UNDERSTAND HOW BEES THINK.«

—

ABOVE

Modern technology makes it increasingly easier to elucidate the highly complex behaviours taking place within a bee colony. Radio frequency identification (RFID) chips are valuable tools, fastened to the backs of bees like miniature backpacks.

—

OPPOSITE

The bee's sense of sight is just as important as its sense of smell outside the dark bee nest. The limits of the visual capacity for orientation can be easily investigated in situations created by the scientist, which are much simpler and more manageable than the case would be in the thick vegetation of a forest – the access point to a feeding station is illustrated here.

—

ABOVE AND ABOVE RIGHT

*Berliner Patrick Kohl capturing a bee
on a flower. The more bees that fly
back and forth between their nest
and the starting point, the easier it is
to perform beelining. The small box
attracts the bees.*

—

BELOW RIGHT

*Patrick Kohl and Benjamin
Rutschmann marking bees, whose
departure path can be determined
by watching them fly by in a process
called beelining*

—

OPPOSITE

*Researcher Benjamin Rutschmann
inspects a tree cavity inhabited by
bees.*

MOVING INTO THE WOODPECKER NEST

When the old queen has left the nest with half of the colony, a temporary camp is set up (see page 121) from which the nest hunters can set off in search of a new living space. These worker bees have very strong preferences with regard to what constitutes a suitable home. The nest cavity must not be too big, since that would make it difficult to regulate the climate; but it should not be too small either, since then there would not be enough room to grow. It should be located high enough, ideally several metres above the ground, and it should be dry. It should also have an appropriately-sized entrance opening which leads into the cavity as high up as possible and preferably faces a favourable direction, north being the least popular.

With these notions in mind, the nest hunters examine each of their respective discoveries painstakingly. American bee researcher Thomas Seeley found that the bees that had discovered potential living quarters even obtained an overview of its dimensions. They do this by circumnavigating its inner surfaces, alternating their route until they have developed a mental image of the cavity's volume. If the scout is convinced that the dwelling is suitable, it returns to the swarm cluster where it makes a case for that destination by performing the waggle dance in front of the onlooking bees. Because each of the nest hunters has generally discovered another option, which they in turn promote by performing a dance of their own, a highly complex series

of behaviours and interactions commences between the bees which ultimately leads the swarm to the best of the new-found abodes.

After decades of in-depth research, Thomas Seeley was able to explain this marvel of communication and teamwork in the bee colony in the most minute detail. At the end of the process, only one destination is chosen for the swarm cluster. That destination is then visited by more and more nest hunters. They fly back and forth between the swarm cluster and the potential nest, possibly marking their route with a pheromone released from the Nasanov gland on their abdomen.

In order to stir up the swarm and incite an explosive departure, all the bees have to raise their body temperature. But how do the bees on the inside of the cluster become aware of the need to do this? What happens is that the bees that

—

The first worker bees from the swarm reach the woodpecker nest chosen by the bee colony as its future dwelling. For some of these bees, it is not the first time that they have visited this cavity. One of the discoverers slowly brought back more and more bees until a small group had gathered together, which then led the swarm to the target.

have already visited the destination and are dancing on the surface of the swarm stop dancing and burrow deep into the swarm cluster, making piping sounds as they go. As they hear the piping, all the bees begin to raise their temperature.

After a few minutes, they are all ready for take-off and the swarm explodes. The swarm cloud, now floating through the air, does not know at first where this journey is taking them. The experienced bees take the helm once more. They buzz back and forth through the cloud of bees like bullets, on an axis in the direction of the objective, until the cloud starts to move in that direction. The experienced bees may also ensure that the target is clearly marked by whirring around the destination and perfuming the air with pheromones from their Nasanov gland.

In the forests of Europe, woodpeckers provide a reliable supply of nest cavities. Germany's largest native woodpecker is the black woodpecker and it builds the most spacious cavities. Abandoned woodpecker roosts undergo changes over the course of time. The tree begins to smooth out the sharp corners of the cavity entrance by growing around it little by little (see page 17). Inside the cavity, degradation processes commence, with fungi and bacteria being the main agents of change. In this manner, the cavity expands, making it even more attractive for the bees than a new woodpecker nest freshly carved into the tree trunk would be. Once the swarm has landed next to the entrance hole to

their new home, they immediately start to move in. The bees gather directly at the entrance and on the inner walls of the cavity around the opening, before wandering up to the ceiling of the cavity and forming a dense cluster. Inside the cluster, they begin building the comb straight away. This bee colony, whose settlement story we followed and documented, erected the combs with the front edges facing the cavity entrance, which is referred to by beekeepers as a horizontal construction. A comb erected parallel to the entrance would be called a vertical construction.

A bee swarm prepares for the move while still in the old nest, with the worker bees getting their wax glands running at full throttle. As a result, they are able to produce a large amount of wax scales immediately upon arrival in the new cavity. The first steps in the comb construction process are impossible to observe as they take place inside a dense ball of bees. What can be seen is that not all the wax scales end up on the construction site, and many of them flutter down to the floor of the cavity. To our surprise, we only noticed bees there occasionally, gathering up this valuable construction material and transporting it to the comb construction site.

After several days of construction work, the first few combs become visible, emerging from between the bodies of the bees – snow-white, semicircular alcoves on the edges of the comb, positioned as close together as possible. It is

—
OVERLEAF

The tip of the swarm reaches their future home. Just a few bees from the colony are able to lead all the others to the destination. Dances on the swarm cluster, flights back and forth between the cluster and the tree, and scent marks at the destination are the key behaviours here - all components that can be found when recruiting bees to a feeding site.

noteworthy that the bees' activities fluctuate significantly depending on the time of day. During the day there is a great deal of movement across the combs, with a loose 'net' of bees hanging under the combs, their legs intertwined with one another. At night, this net becomes denser, extending to the top and covering all the combs. It hangs like a sleeping bag over the entire comb complex. When the temperature outside drops, the mesh is pulled tightly together, and when it is very warm outside the hive, it widens.

It is safe to assume that the construction chains, of multiple insects joining together and hanging like a chain, which can be seen while the bees are building the comb in beekeeper hives are a legacy of net building in a natural bee nest. Currently, there is no definitive answer as to the purpose of this behaviour. Does the 'sleeping bag' serve to regulate the temperature and humidity around the combs? Is it simply a break room for exhausted wax producers and construction bees? Or does it provide protection for the virgin combs?

Once the comb construction is well underway, the interior design of the cavity begins. The walls are not simply left as they are. Two steps are required to make them fit for bees. First, a group of 'washboarding bees' removes any loose particles from the surface of the inner walls. These bees scour the cavity walls relentlessly using their mandibles. The particles removed in the process fall to the cavity floor, creating a preliminary ground cover together with the wax scales that have been dropped. Once the walls have been smoothed, the bees apply a coat of propolis, a resinous mixture created by the bees themselves from material exuded from tree buds; the function of the coating is primarily to act as a protective layer. From what we could see the bees do not just cover gaps and cracks in this substance – since our cavity did not have any – but rather coat the entire inner walls with it, starting at the uppermost combs, which stretch downwards farther and farther as time goes by.

Preparing the cavity walls in this manner undoubtedly helps establish a certain 'feng shui' in the bee dwelling. The propolis coating changes the physical properties of the cavity walls, which in turn affect the bee climate and the air humidity in particular, though recent evidence suggests that high humidity in a nest cavity is not a problem for a honey bee colony. Comparative measurements taken in bee trees and standard beekeeping hives confirm this, with more favourable conditions prevailing in trees. The anti-bacterial and anti-fungal properties of the propolis, which eventually covers the bee nest on all sides like a thin armour, help sustain the good health of the colony.

—

LEFT

Everything starts with a chance discovery. A scout bee has discovered the old, long-since-abandoned woodpecker nest and is inspecting it thoroughly. If she is convinced of the quality of the 'real estate', she will return to the waiting swarm cluster and promote that black woodpecker nest by means of dance. The more the bee likes the object, the more intense its recruitment campaign.

—

OPPOSITE

The swarm moving in. The black woodpecker nest has been uninhabited for some time, the tree has grown around the oval opening. Inside the tree, degradation processes have enlarged the original cavity, especially downwards, which is very appealing to the bees. A fresh woodpecker cavity offers very little in terms of development possibilities.

—

OVERLEAF PAGE 136

The swarm has completely moved into the cavity, first surrounding the entrance hole and then running up the inside walls

—

OVERLEAF PAGE 137

About one and a half hours after moving in, most of the bees gather together to form a cluster around the queen at the top of the cavity.

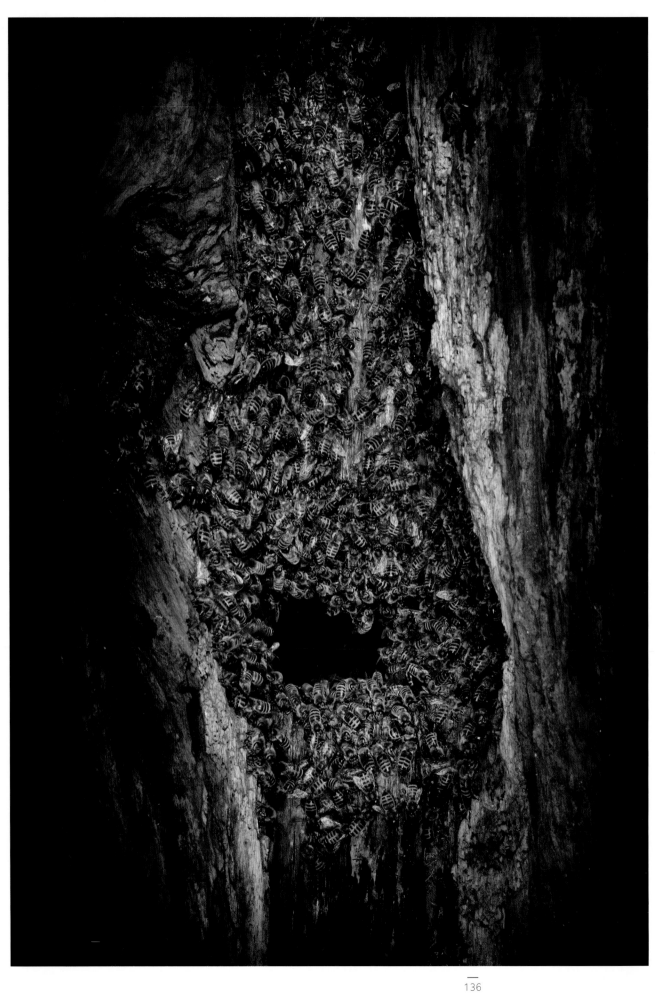

During the first night in the new dwelling, around four in the morning in this case, the bees create a tightly-packed sphere. However calm the bees may appear from the outside, they in fact have an especially active metabolism during this phase. The construction of the combs has to commence quickly. In order to do so, the swarm bees are poised to start producing wax immediately. They left their old nest with full honey stomachs and can now let their wax glands get to work with the material components and energy provided by the honey.

—

RIGHT

The thermal camera shows a glimpse from inside the cavity towards the cold, black entrance hole and the warm ball of bee bodies on the ceiling of the black woodpecker nest. The unusually high generation of heat in a bee cluster without a brood is a sign of the upcoming production of wax.

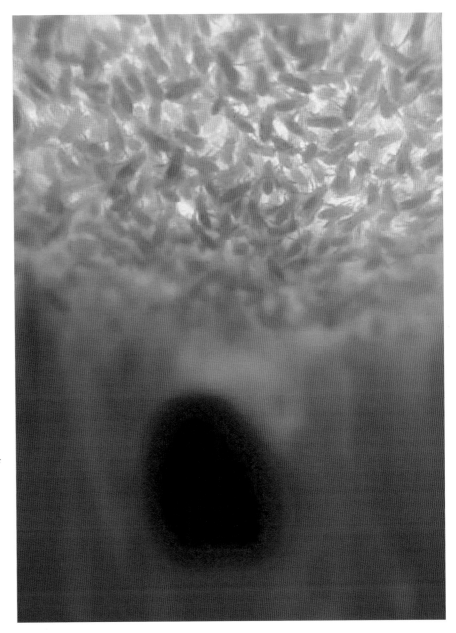

On the first morning after moving in the bees form several dense layers of bodies on the ceiling. For the first time since the move, a 'bag' can be seen around it, formed by bees interlocking their legs with one another. This bag will remain intact in the weeks and months to come, acting as a highly dynamic structure whose 'mesh' can be constricted or enlarged depending on the time of day, conditions and needs. This raises the suspicion that the construction chains formed by bees in hives are a futile attempt at emulating this behaviour. The lack of space in beehives only allows for chains below the comb frames at best, which are construed as construction chains due to their forcibly perpendicular appearance and are intended to specify the direction of the comb construction. A look into a tree cavity casts doubt on this interpretation.

No active task can be discerned
among the bees interlocked with
one another. They do not produce
any wax scales, and they do not
participate in building the combs.
They remain steadily in the same
position, anchored to their respective
neighbours, for many minutes or even
hours at a time.

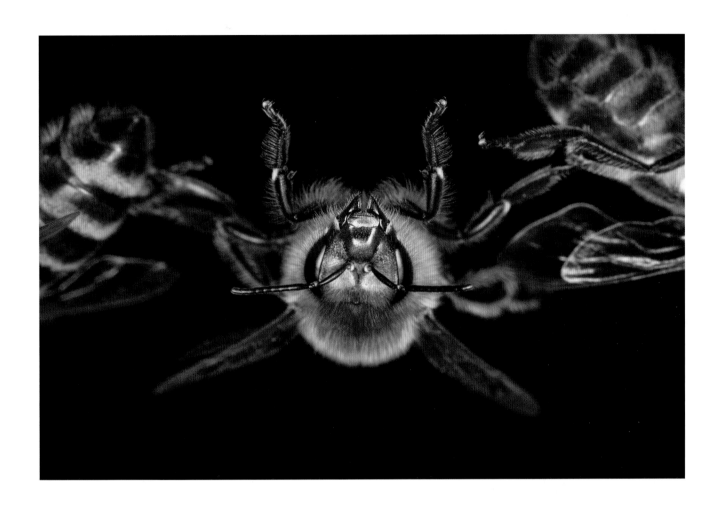

It also does not seem as though the bees forming the 'bag' are sleeping based on the position of their bodies. The feelers and legs of this bee hanging with its stomach pointed up are not dangling downwards, as is customary for sleeping bees resting in the comb alleys.

ABOVE
Occasionally, drones fly with the swarm. When building a new nest, they do not have any tasks and are removed from the cavity.

—

OPPOSITE
This bee is signalling the way into the cavity to the swarm. It is releasing the pheromone geraniol from its fully open Nasanov gland, the glistening yellow field on its abdomen. The wind generated by it wings sends the scented cloud outside.

»SCENT TRAILS LEFT BY THE BEES ARE IMPORTANT MARKERS TO DESTINATIONS OUTSIDE IN THE FIELD AND BACK TO THE NEST.«

—

OVERLEAF
Roughly two weeks after the bees move in, this colony's first combs start to become visible. Construction began right from the outset, but it had remained hidden below the bodies of the bees.

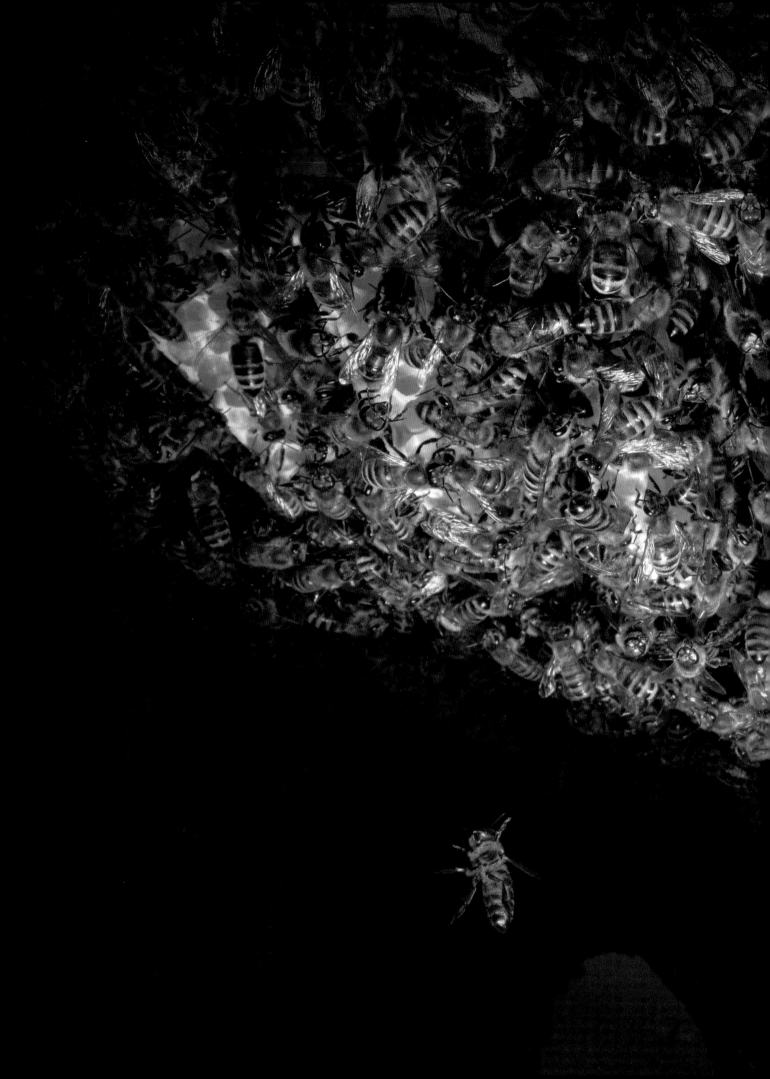

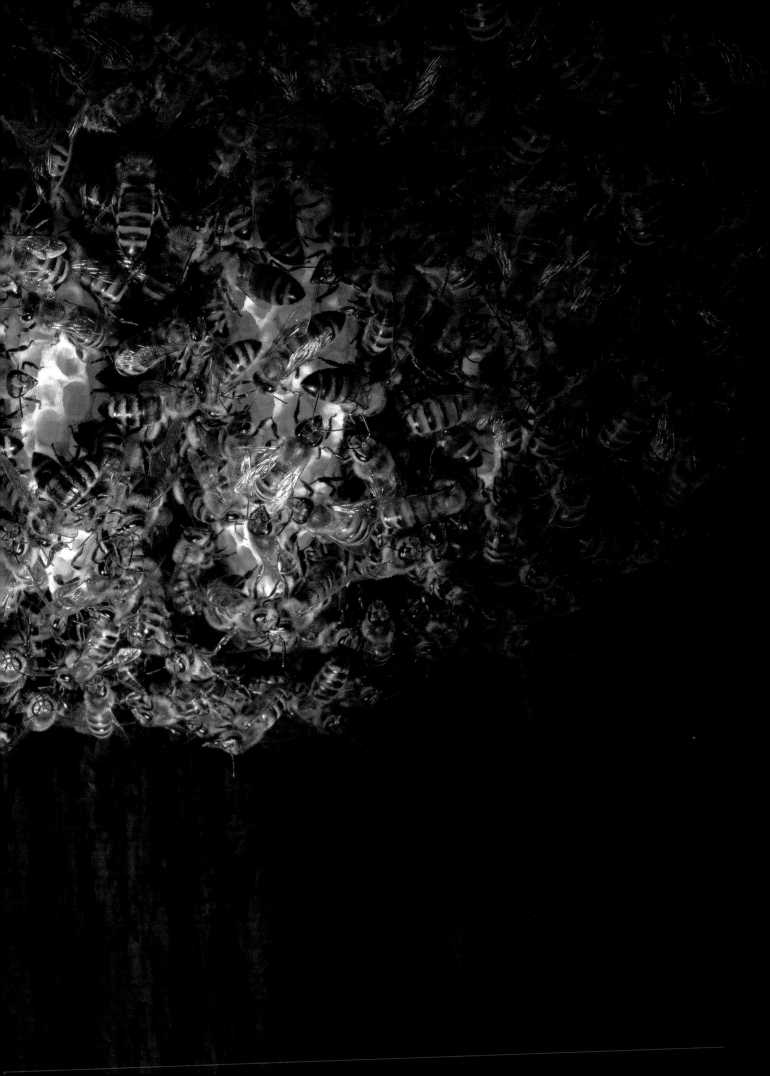

OPPOSITE

The eight wax glands on the abdomen of the swarm bees operate at full capacity. Sweating out wax, chewing up the scales and adding enzymes all take place behind the scenes, unseen by the observer.

—

ABOVE

A construction bee taking wax scales from a nest mate with its mouthparts, the mandibles.

—
OPPOSITE AND TOP AND OVERLEAF
Construction bees chew up fresh flakes of wax and spread them onto the edges of the cells at the borders of the combs. The combs are white initially. They only become discoloured over time, changing to yellow and then dark brown.

—
ABOVE LEFT
More than just a few wax scales are lost during the handover, landing on the floor of the cavity.

—
ABOVE RIGHT
Construction bee collecting lost wax scales.

—
PAGES 154 AND 155
The combs now grow downwards, fastened to the back of the cavity and maintaining a distance from the entrance hole. This is the view during the day (page 154). Typical appearance on a warm summer night. The entrance to the nest is sealed off by bee bodies, with the bee bag hanging loosely below the combs, thus connecting them to one another (page 155).

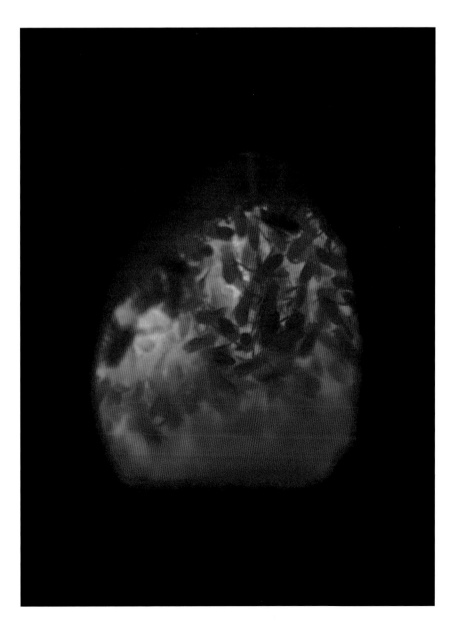

—
OPPOSITE

*The nest entrance seen from
outside at first light after a warm
summer night. A thick plug made
of bees sealed the entrance during
the night. Inside the cavity, the bees
leave many bee-lengths of distance
between the entrance hole and
the combs, with the bee seal only
filling the direct entrance and not
extending into the cavity. On cold
nights, the bees do not form this
plug.*

—
RIGHT

*9:30 pm. A look through a
thermal camera at the cavity
entrance from outside with warm
outdoor temperatures. The bees
are starting to seal off the nest.
Those that have found their
position stay still and allow their
body temperatures to sink (blue
on the image).*

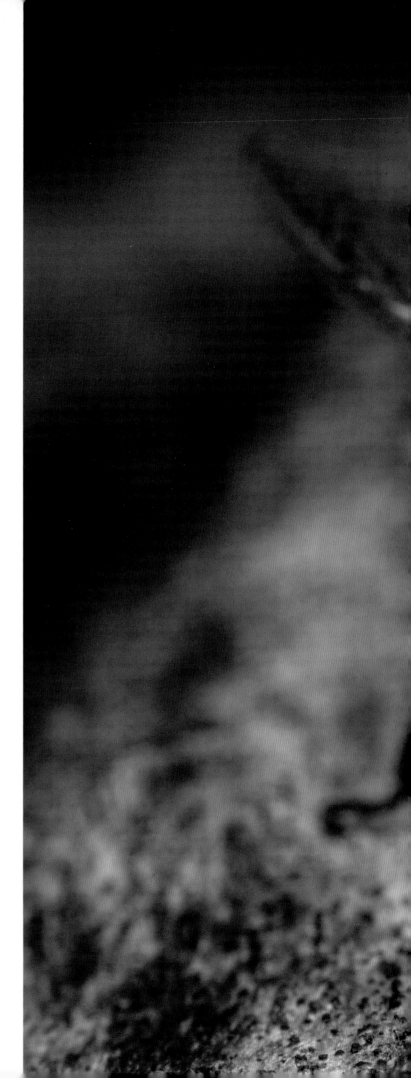

The bee colony is not only occupied
with building the combs after
moving into the black woodpecker
nest. A large number of bees are
busy washboarding the inside walls
of the tree cavity. To do this, the
bees open their mouthparts, the
mandibles, and place them on the
wall, pulling them front to back
towards their bodies across the wood
on the cavity wall. They then lift
their heads, moving their mandibles
forward once again and placing
them back onto the wood. The entire
process occurs rapidly with frequent
repetitions. After a while, the bee
walks a bit farther and works on a
new area. This frees the cavity wall
of loose particles and smooths it
out. Beekeepers are familiar with
and puzzled by washboarding, since
the behaviour seems pointless on
the smoothly planed boards of the
beehives. Studying bees in their
native environment, tree cavities,
sheds light on the purpose of this
innate behaviour.

—

ABOVE
Many eager washboarding bees smooth out the inner wall of the cavity around the entrance.

—

OPPOSITE
The smoothly planed surfaces of the cavity walls are now covered with a 'carpet' of propolis.

—

OVERLEAF
This was previously unknown as well – what is not possible in a hive, occurs readily in the tree cavity. Bees interlocking their legs with one another to form a net below the combs, which spans the entire diameter of the tree cavity and is anchored all the way around the cavity wall. The mesh in this net, shown here from below, can be pulled tightly together or opened widely apart. It is a complete mystery as to which factors have an influence on this. Regulating the climate between the combs by opening and closing off ventilation channels is one possible explanation.

TRADITIONAL BEEKEEPING

Long before German teacher and naturalist Christian Konrad Sprengel first described the pollination of flowering plants in his book *Discovery of the Secret of Nature in the Structure and Fertilization of Flowers* in 1793 – and added honey bees to our collective knowledge as important pollinators – the benefits of honey and wax were well known and valued.

For early humans, roaming the forests as hunters and gatherers, contact with bee colonies was rather coincidental, though honey and wax were surely recognized as much-coveted finds. And the pivotal discovery that a smoking fire placed close to a bee nest reduces the number of painful stings when extracting honey was probably by chance as well.

As Stone Age humans became sedentary, their movements through the forest around their dwellings became more limited and the probability of them stumbling across new bee colonies diminished. Instead, repeated visits to bee nests that had already been discovered probably became the norm. Year after year, they could check whether the bee colonies they were familiar with were still offering a new honey harvest. To do this, the combs were most likely pulled through the openings in the tree cavities in fragments, as can be discerned from the oldest known painting of a Neolithic honey hunter – the 8,000-year-old paintings in Las Cuevas de la Araña, near Valencia, Spain.

From these beginnings, the practice of honey hunting evolved over the centuries with the help of increasingly appropriate tools. The honey hunters gathered honey from wild bee colonies, but no longer from natural bee nests discovered accidentally. Humans created artificial cavities in trees, fashioning them after the bee dwellings they found in nature. The volume of the artificial cavities, their openings and their positioning several metres above the ground made them highly attractive prospects for bee swarms in search of a new home. The honey hunters did not even have to worry about stocking them with bees as the bee swarms moved in on their own, and the tops of the bee trees were often chopped off to reduce the risk of the wind breaking them at the newly created weak point in the trunk. The honey hunters marked the bee trees as their property by carving personal markings into the bark; and so the practice of honey hunting became recognized and legally regulated. These regulations included the right to carry weapons, a very useful privilege that offered them some protection against the other forest dwellers, such as the bears which also preyed on the bees.

—

Polish honey hunter Andrzej Pazura on a bee tree in summer 2019 with traditional equipment. Honey hunters gather honey from wild bee colonies, for which they supply artificial tree cavities – an ancient craft that is still practised to this day.

Honey was the most prized bee product since it was, at the time, the only sweetener available. As a result, honey hunters could barely keep up with demand. Likewise, they were increasingly unable to fulfil the growing demand for wax. Beeswax was becoming more and more important as the key ingredient for the production of candles, which were in demand for illuminating monasteries, churches, castles, and palaces, as well as an ever greater number of city dwellings. As a consequence of this growing demand, the practice of beekeeping emerged in parallel with the lingering tradition of honey hunting.

In contrast to honey hunting, beekeeping considers honey bees as, and treats them like, domestic animals. Beekeepers procured honey bees from the woods and settled their bee colonies in artificial housing, known as beehives — of which a countless number of types, shapes and models have been developed. To ensure that beekeepers could keep an eye on their bee colonies, the hives were rarely located within the forest. At best, in this context, beekeepers would simply move their hives near aphid colonies for a few weeks if they were hoping to collect forest honey.

However, the blossoming of beekeeping did not cause honey hunters to disappear completely. In Eastern Europe especially, this type of beekeeping is still being practised, and honey hunters have not reported even remotely similar disasters in terms of colony collapse disorder as beekeepers. What might be the reason for this?

Honey hunters refrain from interfering with their bee colonies in any way, aside from harvesting honey. They simply let nature run its course in the forest, the native habitat of the bees. The bee colonies live in artificially created tree cavities, with properties much closer to those of natural cavities in trees, than those of hives used in beekeeping. Honey hunters do not actively populate their tree cavities with bees, but rather wait for swarms of bees to stumble upon and move into these dwellings of their own accord. In the artificial tree cavities, a mini ecosystem develops over time similar to that found in a woodpecker nest colonized naturally by bees. The bee colonies propagate by means of swarming and are subject to natural selection. The honey hunters do not take any action to fight bee diseases or parasites.

All of this, combined with the problems beekeepers face with regards to honey bees, has led to a new found interest in the ancient practice of honey hunting. Furthermore, this has sparked an increased interest in wild bee colonies in our forests in general. The resulting studies have yielded several surprises. There are still wild bee colonies living in tree cavities, and they are more numerous than one would have thought possible. For example, in unmanaged German beech forests, Benjamin Rutschmann and Patrick

Kohl found one woodpecker nest inhabited by bees every five square kilometres.

Studies by Thomas Seeley, a pioneer of modern beekeeping with wild bee colonies, have uncovered even higher densities. He found one inhabited bee tree per square kilometre in the Arnot Forest, New York, USA. Furthermore, Seeley established that these wild bee colonies were in no way spared from infestations of the parasitic *Varroa* mite but, unlike bee colonies taken care of by beekeepers, are able to live with this problem. Genetic studies carried out by Seeley and colleagues on wild and 'domestic' bee colonies have revealed significant genetic differences between the colonies, although they were living in the same region. As an explanation for the coexistence between bees and mites, it can be concluded that natural selection has allowed the examined population of bee colonies living in tree cavities,

—
ABOVE
Still used today – a traditional cage for transporting queen bees.
—
OVERLEAF
Traditional and still in use – equipment for hollowing out trees and taking care of bee nests.

and not cared for by beekeepers, to adapt to a life with the *Varroa* mite and survive.

All these fascinating observations have led to a fresh look at the old craft of honey hunting. The fact that honey hunters largely leave the bee colonies to their own devices suggests that there are few, if any, genetic differences between these colonies and their wild counterparts, though scientists are currently researching this.

Will the experiences gathered from the practice of honey hunting, represented as a type of encounter between bees and humans in the forest, some day influence modern beekeeping? Could this encounter, by which bee populations are preserved in unspoiled regions, giving rise to a gene pool formed by natural selection, some day serve as a

valuable genetic resource for beekeepers? And might this type of bee cultivation once again gain a foothold it has not possessed since its heyday, hundreds of years ago, before it was driven out by the practice of beekeeping?

It should come as no surprise if another chapter is added to the story of humans and bees in the forest, with the return to the roots of their first encounter serving as a model. It is something that could only be of benefit to both parties.

ABOVE
Access point to a tree cavity created
by a honey hunter. A wooden wedge
reduces the size of the entrance,
thus making it easier for the bees to
defend themselves against predators.
—
OPPOSITE
View of a honey hunter bee tree.
The white combs were rebuilt not
long ago because the honey hunter
harvested the honey-filled combs
here.

»HONEY HUNTERS LEAVE THE BEES IN THEIR ORIGINAL HABITAT.«

—
ABOVE LEFT
The wing feathers of a dove are softer than commercial brushes. They are used by honey hunters to sweep the bees gently away from the combs after smoking out the bee nest.
—
ABOVE RIGHT
Honeycombs recently broken out of a bee tree.
—
OPPOSITE
Andrzej Pazura tries freshly harvested honey, which his bees produced from flowers and aphids in the forest.

OPPOSITE AND ABOVE

Each honey hunter marks his bee trees with a personal emblem, the signature of two honey hunters in this case - a method of displaying ownership that has been practised by honey hunters from the Middle Ages to today.

OVERLEAF

An alternative to hollowing out living trees is hanging log hives, as shown here in a Polish forest. However, swarming bees in search of a home prefer cavities in trees to log hives. Astonishingly, since they seem as near an archetype as possible to a real tree cavity, even the log hives are perceived by the bees as different from a true tree cavity.

EPILOGUE

Honey bees and beekeepers belong together. This is so firmly rooted in our identity that many people are astonished to hear that there are still bee colonies living in the wild, without the care of any beekeeper. And if an attentive wanderer in the forest chances upon a bee colony, they initially assume that these bees must have swarmed away from a beekeeper.

Honey bees are considered 'domestic' animals which, if not needing the care of a beekeeper to stay alive, would at the very least be worse off without humans. This 'assisted living' for bees started thousands of years ago. Over 4,000-year-old drawings of Egyptian pharaohs illustrate a highly developed form of beekeeping, with cylinders made of dried mud from the Nile, together with the extraction and storage of honey. Archaeological finds in the Middle East reach even further back, with 10,000-year-old fragments of clay pots exhibiting traces of beeswax. Mankind began to settle down during this period following the emergence of agriculture.

In his book *The Lives of Bees*, Thomas Seeley poses an interesting theory as to how the first farmers might have started developing bee colonies. Swarming bees might have moved into empty clay pots or baskets lying unused around the dwellings where people lived. Regardless, honey bees form a part of the early history of farming, just like sheep and goats.

One of the characteristic features of farming is man's attempt to domesticate both plants and animals by means of breeding. Altering the genetic material of a group of organisms through breeding requires control over the process of mating or insemination. The resulting offspring must then be sorted systematically to select the individuals that possess the desired characteristics, or that only exhibit undesirable characteristics to a limited extent, or not at all. Those individuals are then allowed to continue propagating. This is how all of our modern varieties of grains and potatoes, species of dogs and cattle, and many other breeds came about, some of which differ substantially from their original form.

How do honey bees fit into all of this? After all, beekeepers also wish for certain attributes that would be beneficial for practical beekeeping. Is it possible to 'breed out' the tendency of a bee to defend itself by stinging, and instead 'breed in' an increased drive to collect nectar, or to concentrate on specific plants when collecting pollen, or to defend itself against parasites such as the *Varroa* mite? In principle, this should all be possible, since all of these and other traits have a genetic basis that can be altered by means of breeding.

But why, after millennia of beekeeping, are there still no honey bees that differ from their primal form to the extent that the dachshund does from the wolf? And do we actually know

what we are doing when we modify the genetic material of organisms by means of artificial selection?

The interplay between the genes, the resulting traits and the environment is highly complex. If one hopes to breed a particular trait and focuses the artificial selection on acquiring that trait, one still cannot be sure what other genes might be altered in the process and how the interplay between the genes will pan out. But one thing is sure - there will be a narrower genetic spectrum available at the end of the selection process than there was in the original population. And this can come at a price. Alternative forms of a gene, referred to as alleles, that were bred out are lost - aurochs can no longer be bred back from domestic cattle.

Thus far, a series of circumstances has prevented us from beekeeping with a breed of bee that has nothing in common with its ancestral form. The honey bee has remained a wild animal even in the hands of the beekeeper. We can also recognize this based on the fact that bees housed in hives continue to display behaviours such as the forming of construction chains, washboarding and using propolis (which they apply to the inside walls of the hives in clumps), which make much more sense in a tree cavity than in a hive. In the planed, cramped beekeeping hive, the bees can only flourish at a rudimentary level, which is why these behaviours seem pointless and puzzle many observers. Only by studying honey bees in their native environment is it possible for us to understand the world, and needs, of the bees and to learn how they developed through the process of natural selection.

—

The smoker imitates a forest fire and causes the bees to retreat.

The main reason for this 'genetic inertia', and why it seems impossible to achieve any lasting results when it comes to breeding honey bees, is the beekeeper's lack of control over all bee pairings. Although better and better tools have been developed over the last 100 years to artificially inseminate a virgin queen bee, there have to date been no successful breeding results that have asserted themselves to a large degree. All breeding efforts will fail to produce permanent changes in the genetic material of the honey bee population as long as the modified population continues to have access to the outside world.

Bees mate in flight, high in the air – a behaviour that has very rarely been observed and never photographed or captured on film. Only an artificially induced pairing has been documented. In order to achieve it, the queen is fastened to the boom on a spinning carousel. A swarm of drones is then released, which is presumably what would also happen outdoors.

If beekeepers were able to control the breeding process so that only the desired pairings were to take place, then the honey bee would have long since departed from its original form. However, attempts to influence the breeding of honey bees can only be temporarily successful or only possible on a very local basis because, thanks to their mating biology, the bees themselves make sure that their genetic roots are preserved and that they remain, in essence, a wild

animal. These roots reach back through an endless line of generations over millions of years, leading to honey bees in the woods and their dwellings in hollow trees. While we have not altered the honey bee and turned it into a truly domesticated animal dependent on humans in its 'new' form of existence, as is the case with the high-yielding dairy cow, we have radically changed the honey bee's environment and so made it dependent on us.

We offer honey bees artificial dwellings. We move the bee colonies inside these dwellings to locations where the bees can find as many flowers as possible. We provide them with sufficient supplies of food and energy before the winter and wrap up their housing to keep them warm. We fight off bee diseases and parasites. We take care of any emerging bee foes, such as the small hive beetle or the Asian hornet. And this is not to mention the affection, or even love, that a beekeeper has towards his wards.

Our honey bees should be thriving. Strangely enough though, that is not the case. The bees are not doing well, and the beekeepers are not happy. If you take a look at the statistics regarding the number of bee colonies housed in hives, it becomes clear that the number of bee colonies is on the rise. In some areas, there are even so many bees that this has led to stiff competition both between the colonies and with other pollinators, such as solitary bees living in the wild. However, the number of bee colonies has only increased

—
OPPOSITE LEFT
All extremes are close together— bee pasture and bee wasteland.
—
OPPOSITE RIGHT
The extended mouthparts and the body position indicate that this bee was a victim of agrochemistry.
—
LEFT
Considering the forest as an economic factor and a diverse habitat is not necessarily contradictory.

because there are more and more people who are worried about developments with a negative impact on the bees, actively seeking to do something for the animals by keeping bees of their own. Beekeeping is becoming increasingly laborious and burdensome for beekeepers, in terms of both time and money. If the beekeeper does everything right, then there is still a high probability that his bees will be well off. However, the times of Wilhelm Busch, a German humorist, poet, illustrator, painter and bee keeper, whose interventions in his bee colonies only involved extracting honey and trapping swarms, seem long gone.

It is uncertain where our long journey is headed. This depends on the manner in which we practice farming, how we change the climate, and how we deal in the future with the world of living beings, of which we are a part.

The fact that there are still resilient honey bee colonies in our forests, harbouring the potential to some day help us out of dead-end situations resulting from well-intended, but poorly executed developments, should give us cause to take an even greater interest in these bees and to make efforts to conserve both them and their environment.

BEE PHOTOGRAPHY

Ingo Arndt

Spread out over two summers, I spent a total of eight months photographing honey bees. It quickly became clear that I would be focusing on free-living honey bees in our forests. Much is already known about the course of a bee's life under the care of a beekeeper, and has already been captured on film. By taking photos of wild honey bees, however, I would be breaking new ground. Until now, no other photographer has documented this secret life in the forest.

The project required a great deal of effort. It was only possible to snap shots of wild bee colonies at the entrance of their tree cavity by hanging from a rope 20 metres (66 feet) off the ground. For pictures of the honey bees colonizing a former black woodpecker nest, I had to build myself an observation lodge, which would grant me a glimpse into the dwelling of the small flying insects. I spent countless hours capturing the formation of a nest built out in the wild. The hot summers during the last two years regularly turned my observation lodge into an oven. For other shots, extensive technical measures were required. For example, I built flight tunnels equipped with photoelectric sensors, special observation tents in order to film a queen laying eggs, or I set up water points in order to photograph the bees in the act of drinking. As a reward for my painstaking work, I obtained extraordinary pictures.

»WORKING WITH HONEY BEES REQUIRES A CERTAIN CAPACITY FOR SUFFERING.«

—

TOP

Taking photos in the observation lodge with a view into the bee nest.

—

ABOVE

With the camera outside a black woodpecker nest colonized by honey bees.

—

RIGHT

Bee shooting while hanging from a climbing rope, 20 metres (66 feet) above the ground.

All of the images in this book were captured using a high-resolution, 50-megapixel single-lens reflex camera. Various macro and magnifying lenses were required in order to ensure that the photos of the tiny flying insects filled the page. Most of the over 74,000 pictures for this project were taken with the help of various flash attachments.

OPPOSITE ABOVE
Even our backyard was used as a photography studio for honey bees.

OPPOSITE BELOW LEFT
Ingo Arndt photographing honey hunters in the forests of Poland.

OPPOSITE BELOW RIGHT
Flight tunnel for photographing flying honey bees at high-speed.

BIBLIOGRAPHY

Arndt, Ingo & Tautz, Jürgen: *Architektier – Baumeister der Natur*. Knesebeck, Munich 2013.

Kohl, Patrick & Rutschmann, Benjamin: The neglected bee trees: European beech forests as a home for feral honey bee colonies. *PeerJ* 6: e4602; DOI 10.7717/peerj.4602, 2018.

Kohl, Patrick & Rutschmann, Benjamin: Versteckt und unerforscht: Wild lebende Honigbienen in unseren Wäldern. *Deutsches Bienenjournal*, June 2018.

Ruppertshofen, Heinz: *Der summende Wald*. Ehrenwirth, Munich 1972.

Rutschmann, Benjamin, Kohl, Patrick & Roth, Sebastian: Beelining – wie man wild lebende Honigbienen findet. *Deutsches Bienenjournal*, July 2018.

Seeley, Thomas D.: *Following the Wild Bees: The Craft and Science of Beehunting*. S. Fischer Verlag, Frankfurt a. M. 2017.

Seeley, Thomas D.: *The Lives of Bees. The Untold Story of the Honey Bee in the Wild*. Princeton University Press, Princeton 2019.

Tautz, Jürgen (with photographs by Helga R. Heilmann): *The Buzz about Bees*. Spektrum Akademischer Verlag, Heidelberg/Munich 2007.

Tautz, Jürgen & Steen, Diedrich: *The Honey Factory*. Die Wunderwelt der Bienen – eine Betriebsbesichtigung. Gütersloher Verlagshaus, Gütersloh 2017.

Wagner, Max: *Das Zeidelwesen und seine Ordnung im Mittelalter und in der neueren Zeit*. Saxoniabuch, Dresden 2016 (reprint of original edition from 1894).

INGO ARNDT has been considered a prominent wildlife photographer for many years. He is well known for his extensive nature stories, which are regularly published in magazines such as *National Geographic*, *GEO* and *Terra Mater*. To date, he has published 18 books and received numerous awards, including the German Award for Science Photography and two World Press Photo Awards. Ingo Arndt has won Wildlife Photographer of the Year and European Wildlife Photographer of the Year awards on multiple occasions and his photographs are displayed in museum exhibitions all around the world.

THANK YOU

Photographing such a large-scale project is only possible with the help of others. Without their support, many of the pictures in this book would not have been possible. I would therefore like to thank the following people, institutions and companies:

Jacek Adamczewski, Anna Brand, Roland Brand, Petra Diener, Carmen Diessner, Tomasz Dzierzanowski, Marius Jordan, Winfried Jordan, Christoph Koch, Patrick Laurenz Kohl, Ulrich Mergner, Jan Michels, Jörg Müller, Andrzej Pazura, Piotrek Pazura, Hans Pohl, Alex Roosen, Benjamin Rutschmann, Torben Schiffer, Luis Sikora, Simon Thorn and Hartmut Vierle.

Dr Stefan Berg, Institute for Apiculture and Beekeeping, Veitshöchheim and Prof. Dr Bernd Grünewald, The Bee Research Institute, Oberursel.

AC-Foto, Gitzo, Hahnemühle, König Photobags, Manfrotto and Swarovski Optik.

I would like to thank editor-in-chief Robert Sperl and photo editor Isabella Russ for their vote of confidence in hiring me to photograph a story about honey bees for their magazine, which is what kicked off this project.

I would like to thank the entire team at Knesebeck Verlag for their highly creative and trusting cooperation, as always.

I would like to thank my wife for the stunning book design, as well as for her patience and enthusiasm during the very time-consuming photography work for the project. I would also like to thank her for allowing me to convert our backyard into an open-air photography studio for honey bees over the last two years, and for bravely putting up with the associated and unavoidable bee stings.

In particular, I would also like to thank Jürgen Tautz for the friendly collaboration. Without his tireless efforts, help and expertise, this project would not have been possible.

JÜRGEN TAUTZ is a German behavioural researcher, sociobiologist and bee expert. He is professor emeritus in the Biocentre at the Julius-Maximilian University in Würzburg. Since 2004, he has been the founding chairman of the Bee Research Association Würzburg. Jürgen Tautz develops and manages the two environmental education and environmental research projects HOBOS and we4bee. In 2012, he won the renowned Communicator Prize, which is awarded by the German Research Foundation to scientists who have made outstanding strides to communicate their scientific findings to the public.

THANK YOU

Many questions and ideas that were incorporated into this book and that surround the wild honey bee and its way of life are extremely complex and have puzzled many minds. It is a stroke of luck if one is able to exchange views on the matter productively. I would like to thank the following people, listed in alphabetical order by last name, for conversations whose results were integrated into this book as the 'tip of the iceberg':

Antonio Gurliaccio, Patrick Laurenz Kohl, Gaby Laebisch, Moses M. Mrohs, Felix Remter, Miriam Remter, Sebastian Roth, Benjamin Rutschmann, Roland Sachs, Torben Schiffer, Thomas Seeley, Karin Sternberg, Andre Wermelinger.

I want to thank Ingo and Silke Arndt for almost two years of great teamwork, as well as the crew at Knesebeck Verlag for a level of support that one can only wish for as an author. A special thanks to my wife Rosemarie Müller–Tautz for her understanding, saving her partner from the pangs of a bad conscience for spending so much time on yet another project.

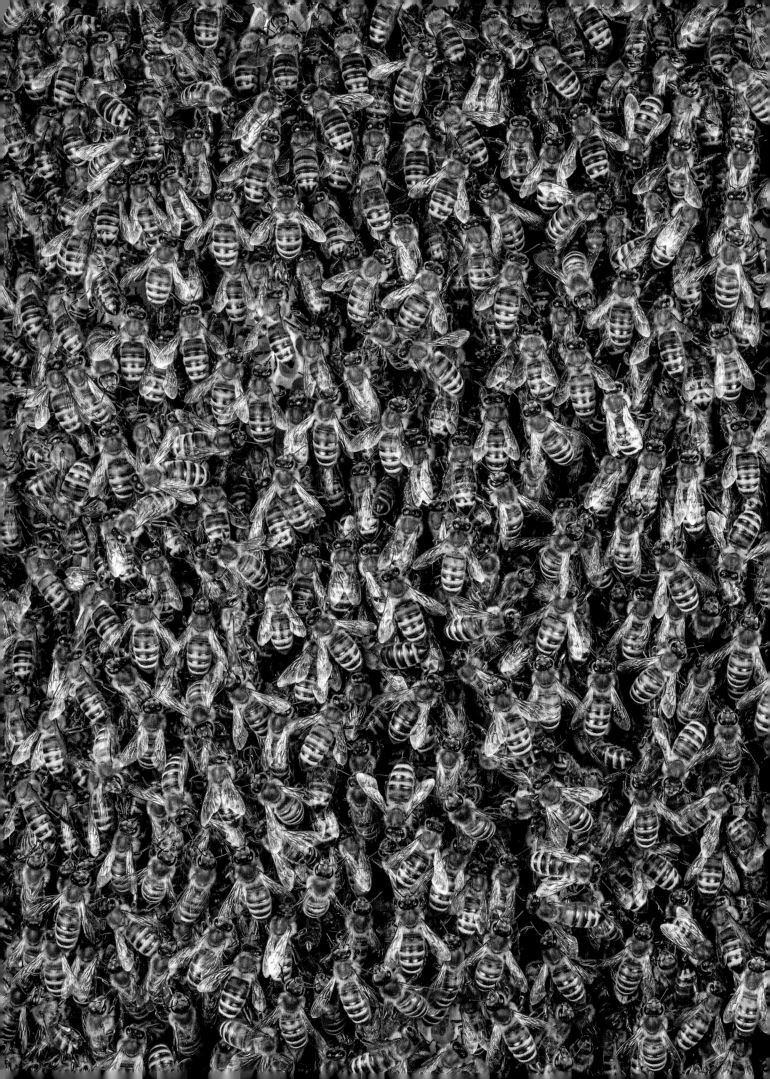

Published in the United States, Canada, South America, Central
America, and the Caribbean in 2022 by Princeton University Press
41 William Street, Princeton, New Jersey 08540
press.princeton.edu

This English edition first published by the Natural History Museum,
Cromwell Road, London SW7 5BD
© The Trustees of the Natural History Museum, London, 2021
Text © Prof. Dr Jürgen Tautz
Text page 5 © Thomas D. Seeley
Text pages 184–186, 188 © Ingo Andt
Photographs © Ingo Arndt, www.ingoarndt.com

Library of Congress Control Number: 2021940021
ISBN: 978-0-691-23508-0
Ebook ISBN: 978-0-691-23509-7

10 9 8 7 6 5 4 3 2 1

Design © Silke Arndt
Printed by Toppan Leefung Printing Limited, China

Originally published by www.knesebeck-verlag.de
Copyright © 2020 Knesebeck GmbH & Co. Verlag KG, Munich
a company of La Martinière Groupe.